T0234121

Kev M. Salikhov

Magnetic Isotope Effect
in Radical Reactions

An Introduction

SpringerWienNewYork

Dr. Kev M. Salikhov
Zavoisky Physical-Technical Institute
Russian Academy of Sciences, Kazan, Russian Federation

This work is subject to copyright.
All rights are reserved, whether the whole or part of the material is concerned, specifically those of translation, reprinting, re-use of illustrations, broadcasting, reproduction by photocopying machines or similar means, and storage in data banks.
© 1996 Springer-Verlag/Wien

Typeset by Zavoisky Physical-Technical Institute, Kazan
Printed by Novographic, Ing. Wolfgang Schmid, A-1230 Wien
Printed on acid-free and chlorine-free bleached paper
Graphic design: Ecke Bonk

With 62 Figures

Library of Congress Cataloging-in-Publication Data

Salikhov, K. M. (Kev Minullinovich)
    Magnetic isotope effect in radical reactions : an introduction /
Kev M. Salikhov.
        p.   cm.
    Includes bibliographical references and index.
    ISBN 3-211-82784-6 (alk. paper)
    1. Free radical reactions.   2. Nuclear spin.   3. Polarization
(Nuclear physics)   I. Title.
QD476.S255   1996
547.1'394—dc20
                                                    95-45887
                                                       CIP

ISBN 3-211-82784-6 Springer-Verlag Wien New York

Preface

During the last 25 years there has been substantial scientific activity in studying spin polarization and magnetic effects in radical reactions. Many scientists have contributed to the development of this field and many remarkable results have been obtained to this date. My involvement in these problems began at the end of 1971 when Prof. Yu. N. Molin asked me to estimate the magnitude of a possible magnetic isotope effect in radical pair reactions. A quick calculation gave a very promising result. At this time we were highly excited about the discovery of chemically induced polarization of nuclear and electron spins. The radical pair model was elaborated to interpret these results. According to this model the spin polarization phenomena originate from singlet-triplet evolution of two unpaired electrons in radical pairs.

Despite the great progress in understanding spin polarization effects it remained questionable whether one could also expect noticeable magnetic effects in radical reactions, i.e., effects of external magnetic fields and magnetic moments of nuclei on chemical reaction rates and product yields. The first demonstration of magnetic effects in liquid-phase radical reactions was obtained in the neighbouring laboratory by Yu. N. Molin, R. Z. Sagdeev and co-workers. We interpreted this result as a field dependence of singlet-triplet mixing in the intermediate radical pairs due to hyperfine interactions of unpaired electrons with magnetic nuclei. Thus, it was also the first demonstration that magnetic moments of nuclei are able to contribute to the rates of chemical reactions not only in principle but quantitatively. This is the physical background for the magnetic isotope effect. We presented these results at a conference in Tallinn (1972) with the result that nobody believed us. Unfortunately at that time we did not know that in 1971 R. G. Lawler and G. T. Evans had obtained similar theoretical estimates of magnetic field and magnetic isotope effects in radical reactions. At the Tallinn conference Gerhard Closs stated in his closing speech that our report would be the most interesting one of the whole conference if indeed proven to be accurate.

Now magnetic effects of up to 10% are well proven phenomena in radical reactions. Here, the magnetic isotope effect is of great fundamental value. This is a new level of our knowledge of the role of nuclear properties in chemical reactions. For a long time it was held that only masses and charges of nuclei play an essential role in chemical reactions. But with the discovery of the magnetic isotope effects it became evident that magnetic moments of nuclei are also very important in chemical reactions.

It was my longstanding wish to write this book. The ideal opportunity arose when the Wissenschaftskolleg zu Berlin (Institute for Advanced Study Berlin) offered me a fellowship for the academic year 92/93. I wish to express my sincere gratitude to the rector as well as to the whole staff at this institution for their invaluable help to accomplish the project within the time limits. I am especially thankful to Kerstin Hoge (Wiko Berlin) for the language editing. Prof. Dr. Dietmar Stehlik's (F. U. Berlin) comments on the text and his availability for discussions were greatly appreciated. I am greatly indebted to Dr. Gerd Buntkowsky (F. U. Berlin) who provided guidance and invaluable support in solving my numerous computer problems. I am pleased to express thanks to Dr. Sergei Akhmin, Dr. Laila Mosina, and Nikolai Berdunov (Kazan Physical-Technical Institute) for their assistance in the final stage of a manuscript preparation. A support of my family, friends and colleagues was a great source of inspiration during my work.

Finally, I am thankful to Prof. N. Turro and the publishers of the journals – Chemical Physics Letters, Tetrahedron Letters and Journal of the American Chemical Society – for the permission to reproduce figures and reaction schemes.

Kev M. Salikhov

Contents

1 Introduction

The isotope effect is a very familiar phenomenon in chemical reactions. The isotope composition of molecules affects their transformation rates in the course of chemical reactions as well as the equilibrium ratio of molecules. Both kinetic and equilibrium isotope effects can originate from the difference between isotopic masses. Isotope substitution changes molecular vibration frequencies, the energy of the molecular ground state (zero point vibration energy), the molecular momentum of inertia, and the effective mass for movement along a reaction coordinate. Atomic masses play a significant role in quantum tunnelling reactions. These isotope mass effects are well-known and are widely used in chemistry for separating isotopes, enriching compounds with definite isotopes, investigating mechanisms of chemical reactions, elucidating structures of transition states and other details of molecular dynamics in the elementary steps of chemical transformations. There are many monographs and reviews discussing comprehensively isotope mass effects (see, e.g., [1–3]).

However, in the last two decades it has been realized that magnetic moments of nuclei can also influence routes and rates of chemical reactions (see, e.g., [4–6]). Isotopes often have different nuclear magnetic moments. For example, the nuclear magnetic moment of deuterium is about four times smaller than that of hydrogen, ^{12}C has zero moment, but ^{13}C has non-zero magnetic moment, etc. Thus, one can expect that the magnetic isotope effect will manifest itself in chemical reactions.

Nuclear magnetic moments can reveal themselves through the hyperfine interactions (hfi) of unpaired (valence) electrons with nuclei. With an isotope substitution the hfi changes according to the nuclear magnetic moments, this means that the hfi is selective to isotopes.

Typically, hyperfine interactions are rather weak compared to thermal energies. Therefore, hfi should not influence the equilibrium ratio of molecules differing in their isotope compositions and, as a result, the equilibrium magnetic isotope effect is not expected to be of importance in chemistry. At the same time a significant kinetic magnetic isotope effect can be expected under favourable conditions. Until now the magnetic isotope effect has been studied extensively for reactions which proceed through radical pair or biradical intermediate states. The hfi does not conserve the electron multiplicity of these intermediates and thus it can finally change rates of chemical conversions.

How does the magnetic isotope effect (MIE) operate in radical reactions? What could be gained using MIE? What problems will be encountered? Qualitatively and shortly these questions will be outlined in the following pages. They will be discussed in more detail in the following chapters.

1.1 Origin of magnetic isotope effect in radical reactions

Many chemical reactions in condensed phases proceed via the formation of radical pairs or biradicals as short-lived intermediates (see, e.g., reviews [4–6]). Radical pairs (RPs) are formed as a result of decomposition of molecules into two radical fragments. Transfer of an electron from donor to acceptor molecules is another possible way to produce RPs, in this case radical-ion pairs. Two radicals, which diffuse in a condensed medium, also pass through an intermediate pair state when they meet accidentally. In a condensed medium these RPs are considered to exist within a "cage" of solvent molecules. The term "cage" will be defined in the next chapter.

Two unpaired electrons of an RP or biradical possess a total spin 0 or 1, so their electronic state may be singlet (S) or triplet (T), respectively. In RPs the unpaired electrons of two partners are well separated, their wave functions overlap very little and the exchange interaction between them is rather small. This means that the two electronic states of RPs – singlet and triplet – are degenerate or quasi-degenerate: electronic singlet and triplet terms practically coincide. Figure 1.1 illustrates this feature of electronic terms of two separated radicals.

One can say that there is a resonance of electronic S and T states in RPs. Under these circumstances even very weak, i.e., minor, interactions can convert RPs from one state to the other, therefore induce

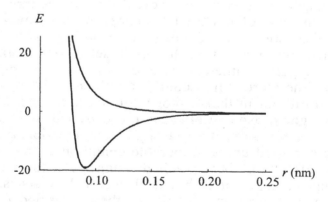

Fig. 1.1. Schematic diagram of the energy (arbitrary units) of RP singlet (S, bottom curve) and triplet (T, top curve) states as a function of the interradical distance r. At moderate distances the two terms approach each other and finally coincide.

intersystem crossing, and change routes of chemical transformations. In other words: minor interactions can result in major consequences.

Qualitatively biradicals are similar to RPs: they also contain two unpaired electrons, interaction between these two electrons is weak compared to the thermal energy associated with the translational motion of the radicals, singlet-triplet dynamics can affect further transformations of biradicals.

However, quantitatively these two kinds of intermediates can result in rather different features. In the case of biradicals the two radical centres are connected via a chain of chemical bonds, so ultimately they cannot separate, and a rather strong exchange interaction can be operative. These properties lead to some peculiarities which can differentiate biradical and RP intermediates.

Several interactions are able to produce S-T conversion in RPs. Among them are: spin-orbit coupling, interactions with external magnetic fields, spin-rotation coupling, hyperfine interaction of electrons with nuclei, interaction of RPs with some paramagnetic additives or other radicals in a system, etc. All these mechanisms of S-T transitions may be in competition. If the hfi mechanism dominates or contributes mainly to the probability of intersystem crossing, MIE will turn out to be the most pronounced.

To get an idea of MIE in radical reactions let us assume that hfi gives the main contribution to S-T transitions in RPs. Organic free radicals very often match this condition. Isotope substitution will change the amplitude of hfi and as a result change the efficiency of S-T conversions in RPs. Due to the specificity of hfi to isotopes the recombination or disproportionation of RPs becomes sensitive to isotope composition.

To visualize the effect of RP singlet-triplet transitions as a cause of MIE let us consider schematically photochemical decomposition of a mixture of fully protonated and fully deuterated organic molecules, MH and MD, respectively (see Fig. 1.2).

Suppose that decomposition has started from an excited triplet state. Then RPs are also created in the triplet state. Typically, the ground state of the molecules, however, is the singlet state. So recombination is allowed only for singlet RPs. In the case considered recombination regenerates starting molecules. In order to recombine and to regenerate the initial molecular state the triplet pair has first to be converted to the reactive singlet state, see Fig. 1.2. The efficiency of the latter process is higher for MH than for MD, since in radicals R_1 and R_2 the hfi with protons is stronger than with deuterons. This will result in a lower quantum yield of photochemical decomposition of protonated molecules compared to deuterated ones. Therefore, the initial mixture will be enriched in protonated molecules, while products of secondary reactions will be enriched in deuterium content. In the same way one can consider MIE in any reaction where a spin-dependent process of radical recombination is in-

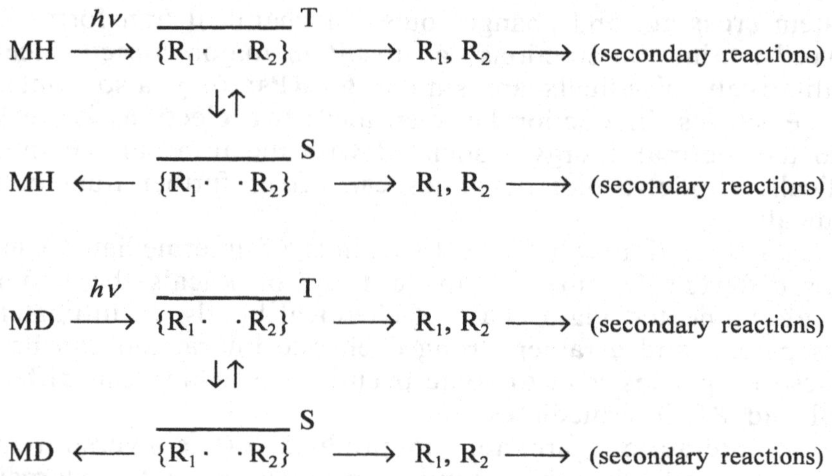

$$\text{MH} \xrightarrow{h\nu} \{R_1\cdot \quad \cdot R_2\} \underline{\qquad\qquad} T \longrightarrow R_1, R_2 \longrightarrow \text{(secondary reactions)}$$

$$\downarrow\uparrow$$

$$\text{MH} \longleftarrow \{R_1\cdot \quad \cdot R_2\} \underline{\qquad\qquad} S \longrightarrow R_1, R_2 \longrightarrow \text{(secondary reactions)}$$

$$\text{MD} \xrightarrow{h\nu} \{R_1\cdot \quad \cdot R_2\} \underline{\qquad\qquad} T \longrightarrow R_1, R_2 \longrightarrow \text{(secondary reactions)}$$

$$\downarrow\uparrow$$

$$\text{MD} \longleftarrow \{R_1\cdot \quad \cdot R_2\} \underline{\qquad\qquad} S \longrightarrow R_1, R_2 \longrightarrow \text{(secondary reactions)}$$

Fig. 1.2. Schematic representation of a photo-decomposition of protonated molecule (MH) and its deuterated (MD) counterpart. For MIE the crucial feature is the formation of the intermediate radical pair state $(R_1\cdot \;\cdot R_2)$: the hyperfine interaction of unpaired electrons with nuclei induces singlet-triplet (S-T) transitions, changing populations of RPs between the reactive S state and the non-reactive T state; the efficiency of S-T conversion strongly depends on the isotope composition of the molecules.

volved. Basically analogous arguments can be put forward as well for reactions in which biradicals are intermediates.

From Fig. 1.2 it follows immediately that the sign of MIE depends on the multiplicity of the RP's precursors. Indeed, when molecules decompose from the singlet state, RPs start from the singlet state as well, and the S-T transitions will deplete the reactive singlet state of RPs. So in this case deuterated molecules have a better chance to regenerate in contrast to protonated molecules which were favoured in the previous example when a triplet precursor of the RPs was considered.

MIE originates from S-T transitions in intermediate RP or biradical states induced by hfi of electrons with nuclei which in turn depend on the isotope composition of the reactants. Generally speaking, RPs and biradicals have different reactivities in singlet and triplet states, therefore S-T transitions do play an important role in kinetics. If RPs and biradicals would have identical reactivity in S and T states then S-T transitions would not manifest themselves.

1.2 Favourable conditions for magnetic isotope effect

We shall see later that, quantitatively, MIE depends on many factors and parameters specifying the mechanism of a reaction under study, mobility of reactants, chemical transformations of radicals, and the S-T evolution in RP or biradical intermediates. But even without going into a detailed description of the MIE phenomenon one can formulate some general features of reacting systems which are promising to display MIE. Some of them follow straightforwardly from the reaction scheme (see Fig. 1.2) and are rather evident.

Different reactivity of singlet and triplet states of radical pairs. Two (or maybe more in some cases) electronic states of intermediates should possess different reactivity. Normally two radicals will recombine or disproportionate preferably from a singlet state rather than from a triplet one. So the RP model matches the requirement that S and T states should have different reactivity. At a first glance it seems that the larger the difference between the reactivities in the S and T states the more pronounced the MIE. But this statement can sometimes be violated. One of the physical reasons for this may be an effect of the chemical reaction on the S-T evolution: chemical transformations of RPs do not disturb spin dynamics only in a case when S and T states do perform the same reactions with exactly the same rates.

It seems reasonable to expect that MIE will be more pronounced if RPs (biradicals) start from a state which is less reactive, normally it will be the triplet state. Then the product formation might be controlled by $T \rightarrow S$ conversion.

Degeneracy of electronic terms. Electronic terms of intermediates should be degenerate or quasi-degenerate. This enables comparatively weak hyperfine interactions to mix essentially different electronic states. This condition may become greatly significant in the case of biradicals with a short distance between radical centres: Here, a strong exchange interaction can decrease the efficiency of S-T transitions induced by hfi. From this point of view biradicals are likely to be less promising to show a large MIE as compared to RPs.

To a certain degree degeneracy of RP electron spin states can be enforced by application of external magnetic fields. This makes it possible to manipulate MIE.

Comparatively long-lived radical pairs or biradicals. The lifetime, τ, of intermediates must have optimal value. On the one hand, it should not be too short, because otherwise the hfi will not succeed to produce S-T conversions or will not reveal itself at all. Let us introduce the hf coupling constant, a, which characterizes the hfi strength. Then we can say that S-T transitions will be efficient under the following condition

$$a\tau \geq 1. \qquad (1.1)$$

Typically, in organic free radicals a has a value about 10^8–10^9 rad/s. Therefore, the lifetime of RPs or biradicals must be in the nanosecond region or larger in order to promote efficient S-T mixing. On the other hand, for the purpose of MIE, the RP lifetime should not be too long either. MIE may diminish when the RP lifetime becomes very long. There are at least two reasons for that. First, not only hfi but also all other possible mechanisms of S-T conversion will operate at longtime intervals and eventually populations of S and T states will be equilibrated, i.e., RPs will be in S and T states with probabilities 1/4 and 3/4, respectively, independent of their isotope composition. For example, in hydrocarbon free radicals electron spins reach their equilibrium in liquid solutions in microseconds. Secondly, the biggest MIE is expected under the following conditions: in RPs with one isotope composition the hfi, which can be characterized by the coupling parameter a_1, does not mix efficiently S and T states because $a_1 \tau < 1$, but after isotope substitution the hfi, which can be characterized by the coupling parameter a_2, does mix them because the opposite condition $a_2 \tau > 1$ is fulfilled. These considerations exemplify that from the point of view of MIE optimal lifetimes of RPs and biradicals are required. According to the above arguments a very crude estimate of the optimal lifetime for liquid phase organic radical reactions is of the order of magnitude

$$\tau \sim 10^{-8} \text{ s} . \tag{1.2}$$

τ can be modified by the viscosity of a solvent or by varying the concentration of radical acceptors which shorten τ. For neutral radicals τ has the order of magnitude [5]

$$\tau \sim \frac{b^2}{D} , \tag{1.3}$$

where b is the reaction radius and D is the coefficient of the mutual diffusion of two RP partners. In the case of non-viscous solvents, e.g., water, Eq. (1.3) gives 10^{-9}–10^{-10} s which is of course rather small compared to an optimal estimate given by Eq. (1.2). Thus, one has to use more viscous solvents with a viscosity in the range 10–100 cP to find a significant MIE. A very promising way is to place reactants within some restricted regions. Many striking results were obtained using micellar systems [6]. An interesting and practically important example of rather long-lived RPs are ionic RPs when a "cage" is determined by the Onsager radius, which is much bigger than a "cage" for neutral partners [7].

Big change of the hyperfine interaction after an isotope substitution. MIE will be more pronounced if an isotope substitution changes the scale of the hfi in radicals remarkably. For example, a substitution of ^{12}C by ^{13}C in the CO group of the radical $C_6H_5CH_2CO$ increases the hfi energy about six times, while a substitution of ^{12}C by ^{13}C in the ring part of the same radical will change the total hfi scale very little. Thus, for carbon isotope substitution in the CO group, one can expect a larger MIE as compared to a case of ring carbons substitution, etc. In designing any experiment on MIE, one has to check data on hfi of those radicals which are expected as potential intermediates. These data are available from electron paramagnetic resonance experiments, Overhauser nuclear polarization experiments, etc.

Dominant role of the hyperfine interaction in radical pair spin dynamics. To reveal a significant MIE the hfi mechanism of intersystem crossing should dominate over other mechanisms, hfi should not be masked by spin-orbit and spin-rotation couplings, etc.

This point requires a careful analysis in each particular reacting system. It is trivial to formulate this latter condition for MIE, but unfortunately it is rather difficult to control contributions from different sources to S-T evolution of RPs and biradicals in an actual experiment.

Favourable conditions for observing MIE as just outlined are probably highly qualitative, but they can be very instructive for experimental design and interpretation of experimental results.

1.3 Some specific features of magnetic isotope effect

MIE always operates in common with an isotope mass effect. So it is important to identify specific features of MIE which will allow us to distinguish the effects of these two different origins. Fortunately there are even some qualitative differences between the mass and the magnetic isotope effects.

An isotope effect behaviour for isotopic triad sequences. One can discriminate mass and magnetic effects unambiguously when there are isotopic triad sequences like $\{^1H, {}^2H, {}^3H\}$, $\{^{12}C, {}^{13}C, {}^{14}C\}$, $\{^{16}O, {}^{17}O, {}^{18}O\}$, etc. To see how it works let us take one of these elements, e.g., oxygen. The mass isotope effect for a substitution of ^{17}O by ^{18}O should have the same sign and practically the same value as the effect in the substitution of ^{16}O by ^{17}O, since masses of isotopes are changing monotonously in the sequence ^{16}O, ^{17}O, ^{18}O. However magnetic moments of these isotopes are varying non-monotonously: only ^{17}O isotope possesses non-zero magnetic moment. Thus for the substitution of ^{17}O by ^{18}O the MIE should have an opposite sign compared to the substitution of ^{16}O by ^{17}O. The same kind of arguments might be applied to any other isotope triad sequences.

Resonance dependence of magnetic isotope effect on microwave pumping. Irradiation of a microwave (MW) field can affect reacting systems in two different ways.

First, it can heat the system due to non-resonant dielectric losses. The mass isotope effect is temperature dependent so MW induced heating could influence an isotope effect.

Another possibility for an MW field effect is associated with electron paramagnetic resonance (EPR) transitions in intermediate radicals. Indeed, there are three sublevels of triplet states and the hfi mixes them non-equally with the singlet state. In the presence of a high external magnetic field, e.g., the triplet sublevels correspond to three possible projections of the total electron spin of RPs on the quantization axis z: T_{+1}, T_0, T_{-1}. The corresponding RP energy diagram is depicted in Fig. 1.3.

From Fig. 1.3, it appears to be evident that MW field induced transitions between triplet sublevels can also change the total efficiency of S-T mixing. This effect of the MW field depends on the frequency of the field in resonance manner. The most noticeable effects will occur when the frequency of the external MW will match the resonance condition for EPR transitions in RPs. Under isotope substitution, the hyperfine splitting of energy levels changes as well as the hyperfine structure of the EPR spectra. Therefore, in a mixture of RPs with different isotope composition one can reach RPs selectively by pumping EPR transitions in RPs with the particular isotope composition only. From this consideration it follows that with MW irradiation one can change the MIE specifically. The most important property of this MW induced MIE is a resonance behaviour when the frequency of MW field is varied or when EPR frequencies are swept with the external magnetic field.

An interesting aspect of the MW effect on RP recombination is its selectivity to the spatial position of nuclei in radicals because the hfi strongly depends on nuclei position. So by MW pumping, one can selectively regulate the so-called primary, secondary, etc. isotope effects.

Dependence of magnetic isotope effect on an external magnetic field. Another peculiar feature of MIE is its potentially strong dependence on the strength of an external magnetic field. As visualized in Fig. 1.3, an external field splits triplet sublevels of RPs. Thus, the efficiency of S-T transitions induced by hfi and MIE will be affected by field variations. Unfortunately, the pattern of an MIE field dependence will not be a unique one, it can be either monotonous or it can display some extrema which correspond to the optimal values of an external field strength. But in any case field dependence of the isotope effect serves as strong evidence that one deals with MIE rather than mass

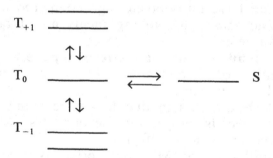

Fig. 1.3. A scheme of energy levels for an RP with one nucleus possessing spin $I = 1/2$. In the high field case, Zeeman interaction of unpaired electrons with the external field splits triplet sublevels. In addition hfi splits T_{+1} and T_{-1} triplet sublevels. The MW field induces transitions which are shown by vertical arrows. Hfi induces S-T_0 transitions.

isotope effect. This latter feature of MIE can be used as a practical application in studying MIE.

Dependence of magnetic isotope effect on multiplicity of radical pair's precursors. When discussing the origin of MIE in Sect. 1.1, it was already pointed out that the sign of MIE depends on the multiplicity of RP's precursors. However, generally speaking, it is not excluded that a mass isotope effect may change sign for the same conditions. Indeed, changing of the multiplicity of molecules is able to result in large differences in the reaction routes. This means that the dependence on the multiplicity of RP precursors should not be utilized for discriminating MIE and mass isotope effects alone but in combination with other features of MIE. In fact this statement is valid for all other features of MIE as well.

1.4 What can be gained from magnetic isotope effect?

It is clear that MIE is a phenomenon of fundamental importance. It connects magnetic moments of nuclei, associated with their spins, with chemical reactivity. Note that nuclear spins themselves sometimes play an important role in chemistry. A classical textbook example is *ortho*- and *para*-hydrogen. But then the question arises how to use this fundamental phenomenon. Generally speaking, it is very easy to answer this question: MIE in chemical reactions can be used in the same way as familiar mass isotope effects either for technological purposes or as a specific method, a unique tool to investigate chemical reactions.

Theoretical calculations and the experimental data available prove that the probabilities of recombination of RPs differing in their isotope composition might diverge up to about 10% which is a rather large isotope effect. Thus, it is definitely feasible to use MIE in technology for isotope separation and isotope enrichment. It may be an especially interesting approach in the case of heavy elements where the relative variations of the isotopic masses become very small while their nuclear magnetic moments can differ very much. Two oxygen isotopes, ^{16}O and ^{17}O, may serve as a good example to demonstrate the advantages of MIE.

Today selective laser excitation is successfully used for isotope selection and enrichment (see, e.g., [8]). Normally, this procedure exploits the isotopic mass dependence of molecular vibration frequencies, basically a mass isotope effect. But in laser stimulated reactions one could also exploit isotopic dependence of the hyperfine structure of energy levels of radicals or paramagnetic complexes to selectively excite particles with a given isotope. Moreover, it would be possible to induce nuclear spin polarization in reaction products of radicals or paramagnetic complexes. Note that for this kind of laser induced nuclear spin polarization it would not be necessary to pass through RP states.

There are very optimistic expectations concerning the applications of MIE to study the mechanisms of chemical reactions in detail. Several examples will be presented later on. Using MIE one can decide whether the reaction under study is proceeding through radical pair or biradical intermediates. A multiplicity of electronic states of RP's precursors and a quantum yield of photochemical decomposition of molecules can be determined with a reasonable accuracy.

1.5 Basic steps in studying magnetic isotope effect

A crucial step in the development of all spin polarization and magnetic effects in radical reactions was the discovery of chemically induced polarization of nuclear spins and its interpretation in the framework of the radical pair model (see, e.g., [4, 5]). In principle, one can readily predict MIE on the basis of this radical pair model. But as it was mentioned already in the preface, the first steps in studying magnetic effects in radical reactions were not trivial, and they required some efforts.

In 1969 B. Brocklehurst [9] suggested that the electron paramagnetic relaxation of ion-radical pairs can affect the ratio of singlet and triplet recombination products. This relaxation mechanism of singlet-triplet transitions in RPs is capable to produce MIE because the anisotropic hyperfine interaction is often substantially contributing to radical paramagnetic relaxation. In 1971 R. Lawler and G. Evans [10] have shown that isotropic hyperfine interaction can lead to a pronounced dependence of RP recombination probability on nuclear magnetic moments. The first experimental demonstration of the significant role of nuclear magnetic moments in ruling out radical reactions in solutions was presented by R. Sagdeev et al. in 1972 [11]. It was shown that product yields for the reaction of pentafluorobenzyl chloride with C_4H_9Li in hexane are dependent on the strength of an external magnetic field. The physical origin of the effect observed is the field dependence of the singlet-triplet transitions in radical pairs induced by the hyperfine interaction. Later, B. Brocklehurst et al. [12], R. Haberkorn, M. Michel-Beyerle and coworkers and K. Schulten, A. Weller and coworkers [13, 14] have demonstrated the substantial contribution of the isotropic hfi and nuclear magnetic moments to the efficiency of a recombination of the ion-radical pairs generated by ionizing radiation and by light, respectively.

Finally, in 1976 two research groups have presented simultaneously and independently their direct observations of MIE in liquid phase radical reactions. A. Buchachenko et al. [15] have succeeded to detect MIE from carbon nuclei (^{12}C and ^{13}C) in the photochemical decomposition of dibenzyl ketone. Yu. Molin and R. Sagdeev [16] were able to register MIE in carbon nuclei as well in the triplet photosensitized photolysis of benzoyl peroxide.

Since that time great efforts were undertaken in studying MIE. N. Turro and B. Kraeutler [6, 17] have demonstrated brilliant magnetic isotope effects in micellar solutions. Their results concerning photochemical decomposition of dibenzyl ketone and its derivatives are able to serve as a textbook example of MIE in chemical reactions. A. Buchachenko and I. Khudyakov [18] have extended the MIE investigations to heavy elements (e.g., U). The first observation of the isotope effects on the reaction rates in micellar solutions was made by S. Nagakura, H. Hayashi and Y. Sakaguchi (see [19]).

MIE is now a well established phenomenon. Maybe it is time to think also about the technological applications of MIE. To this end it would be highly desirable that chemists will realize and accept basic ideas of MIE, and the theoretical backgrounds of the MIE phenomenon.

2 Main concepts of the theory of magnetic isotope effect

This chapter is aimed to discuss the theoretical background of MIE in more detail. We will present the basic concepts and main parameters exploited in the applications of MIE. When analyzing experimental data, the observables should be expressed in terms of the effect of the isotope substitution on the recombination of intermediate radical pairs (RPs) or biradicals. Therefore, the chapter starts with a discussion of the relations between observables and RP's recombination probabilities for some types of reactions. However, the focus of the chapter is the concept of radical pair and singlet-triplet transitions in RPs induced by hyperfine interaction. The illustrations will demonstrate how singlet-triplet transitions change after an isotope substitution.

2.1 Macroscopic and microscopic parameters

The microscopic parameter which is directly connected with MIE is the probability of the recombination of RPs or biradicals. It is exactly this parameter which is sensitive to the isotope composition of radicals. A wide variety of macroscopic parameters can be selected to characterize MIE. One possible approach is to measure the concentrations of reactant and/or product molecules differing in their isotope composition. In many cases, the ratio of the concentrations of molecules with different isotope content appears as an experimentally measured quantity. It is not the goal of this monograph to discuss how concentration of reactants, products and intermediates can be determined experimentally. In fact there are many experimental methods. Among them are mass spectroscopic methods, optical spectroscopic methods, nuclear magnetic resonance and electron paramagnetic resonance methods, etc. Some of them will be mentioned when discussing the experimental results.

The relation between microscopic and macroscopic parameters characterizing MIE depends strongly on the particular mechanism of a chemical reaction and the specific experimental conditions. Let us consider a few examples here.

Consider, e.g., a photochemical decomposition of dibenzyl ketone (DBK). A kinetic scheme of this reaction is depicted in Fig. 2.1.

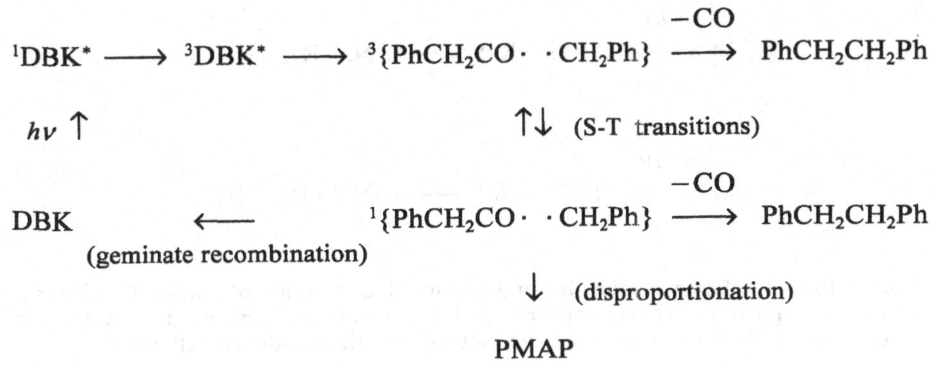

$$^1\text{DBK}^* \longrightarrow \,^3\text{DBK}^* \longrightarrow \,^3\{\text{PhCH}_2\text{CO}\cdot\;\cdot\text{CH}_2\text{Ph}\} \overset{-\text{CO}}{\longrightarrow} \text{PhCH}_2\text{CH}_2\text{Ph}$$

$hv\;\uparrow$　　　　　　　　　　　　　　　$\uparrow\downarrow$ (S-T transitions)

$$\text{DBK} \overset{}{\longleftarrow} \,^1\{\text{PhCH}_2\text{CO}\cdot\;\cdot\text{CH}_2\text{Ph}\} \overset{-\text{CO}}{\longrightarrow} \text{PhCH}_2\text{CH}_2\text{Ph}$$

(geminate recombination)

$\downarrow$ (disproportionation)

PMAP

Fig. 2.1. The scheme of the photochemical decomposition of dibenzyl ketone [20]. Here DBK $\equiv$ $(\text{C}_6\text{H}_5\text{CH}_2\text{COCH}_2\text{C}_6\text{H}_5)$ and PMAP $\equiv$ $(\text{C}_6\text{H}_5\text{CH}_2\text{CO}-\text{C}_6\text{H}_4-\text{CH}_3)$.

Excited triplet molecules dissociate and form pairs of radicals which inherit the spin state of their precursor molecules and initially have the total spin 1. However, this triplet state is not the eigenstate of electron spins in RPs. RPs undergo intersystem crossing to the singlet state due to the electron-nuclear hfi and the difference of Zeeman frequencies of two unpaired electron spins. Singlet RPs are able to recombine regenerating the original DBK molecules or they can give the product of disproportionation, PMAP (see Fig. 2.1). Radicals which escape the "cage" are able to give the same products, DBK and PMAP, as well. However due to the decarbonylation the ketyl radical PhCH_2CO is converted rather fast to the radical PhCH_2, so finally the $\text{PhCH}_2\text{CH}_2\text{Ph}$ molecules are formed mainly as a result of the recombination of radicals in the bulk of a solution [20].

The photochemical decomposition of benzoyl peroxide might serve as an example where the recombination of RPs in a "cage" does not regenerate the starting material, the product of the in-cage recombina-

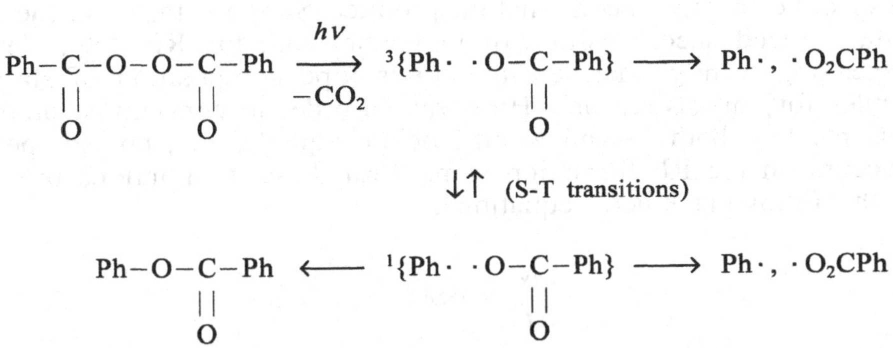

$$\underset{\underset{\text{O}}{||}}{\text{Ph}-\text{C}-\text{O}-\text{O}-\underset{\underset{\text{O}}{||}}{\text{C}}-\text{Ph}} \overset{hv}{\underset{-\text{CO}_2}{\longrightarrow}} \,^3\{\text{Ph}\cdot\;\cdot\text{O}-\underset{\underset{\text{O}}{||}}{\text{C}}-\text{Ph}\} \longrightarrow \text{Ph}\cdot,\;\cdot\text{O}_2\text{CPh}$$

$\downarrow\uparrow$ (S-T transitions)

$$\text{Ph}-\text{O}-\underset{\underset{\text{O}}{||}}{\text{C}}-\text{Ph} \longleftarrow \,^1\{\text{Ph}\cdot\;\cdot\text{O}-\underset{\underset{\text{O}}{||}}{\text{C}}-\text{Ph}\} \longrightarrow \text{Ph}\cdot,\;\cdot\text{O}_2\text{CPh}$$

Fig. 2.2. Schematic representation of the photochemical decomposition of benzoyl peroxide.

$$M \quad \xrightarrow[v]{h\nu} \quad \{R_1 \cdot \quad \cdot R_2\} \quad \longrightarrow \quad N, \cdot R_1, \cdot R_2$$

$$M^* \quad \xrightarrow[v]{h\nu} \quad \{R_1^* \cdot \quad \cdot R_2\} \quad \longrightarrow \quad N^*, \cdot R_1^*, \cdot R_2$$

Fig. 2.3. A formal scheme of the decomposition of a mixture of molecules differing in their isotope composition. The isotope effect for the product concentration N (N^*) is the object of study here. v denotes the rate of the light induced RP formation.

tion is phenylbenzoate $C_6H_5CO_2C_6H_5$ (see Fig. 2.2). The formation of this product in bulk reactions can be strongly reduced by adding acceptors of radicals, so an in-cage recombination can be the dominating channel of the phenylbenzoate formation [21].

Keeping these reaction schemes in mind, let us start with the simplest case to establish the relation between macroscopic and microscopic MIE parameters. Suppose molecules dissociate, giving two radicals which then give products in secondary reactions (Fig. 2.3).

There exists an evolving intermediate RP state when the two radicals are in a "cage" before they recombine or disproportionate. The in-cage recombination either regenerates the starting material or produces new molecules. Let us now investigate MIE for these new molecules N or N^*. This model situation is also seen in Fig. 2.3. Let us consider a mixture of molecules M and M^* differing only in their isotope composition. As an example we assume that molecules without an asterisk have negligible hfi, while radicals marked with an asterisk possess pronounced hfi with the respective magnetic nucleus.

If we define M (or M^*) as the concentration of starting molecules which are not labelled by magnetic isotopes (or labelled by magnetic isotopes), respectively, then N (or N^*) characterize the concentration of the respective in-cage recombination product. Suppose that v is the rate of light induced decomposition of molecules into the RP state. Generally speaking, v might be sensitive to isotope composition of starting molecules for various reasons. However, in order to concentrate on MIE we assume that both M and M^* dissociate equally, i.e., no isotope effect occurs on the RP formation step. With these assumptions one can write the following kinetic equations:

$$\frac{dN}{dt} = vpM \ ,$$

$$\frac{dN^*}{dt} = vp^*M^* \ , \tag{2.1}$$

where p (or p^*) is the probability that RPs, not marked (or marked) with magnetic isotopes, will produce a particular product N (or N*). At the initial stage, the consumption of starting material in the reaction is negligible, so $M = M(0)$ and $M^* = M^*(0)$. Then, Eq. (2.1) results in an evident relation between the microscopic parameters, p and p^*, and the macroscopic observables, the concentration of products, respectively,

$$\frac{N^* / N}{M^*(0) / M(0)} = \frac{p^*}{p} \; . \tag{2.2}$$

An interesting feature of this limit case is that the degree of isotope enrichment does not depend on the degree of chemical conversion M $\rightarrow$ N (M* $\rightarrow$ N*).

Let us now analyze the isotope enrichment of starting molecules in the case when they are partially regenerated through in-cage recombination of RPs. The photochemical decomposition of DBK can serve as an example (see Figs. 2.1 and 2.3). In the case of DBK molecules M and M* may differ, e.g., by their ^{12}C and ^{13}C content. We further assume that the regeneration of DBK via secondary reactions is eliminated by adding acceptors of radicals. The following kinetic equations for the decay of molecules differing in their isotope composition can be written

$$\frac{d M}{d t} = -v(1 - p) M \; ,$$

$$\frac{d M^*}{d t} = -v(1 - p^*) M^* \; . \tag{2.3}$$

Again, we have neglected a possible dependence of the decomposition rate v on the isotope composition of a molecule. Solving Eqs. (2.3) we obtain

$$\frac{M^*(t)}{M(t)} = \frac{M^*(0)}{M(0)} \exp\big(v(p^* - p)t\big) \; ,$$

$$\frac{M^*(t)}{M(t)} = \frac{M^*(0)}{M(0)} \left(\frac{M(t)}{M(0)} \right)^{\frac{p-p^*}{1-p}} \; . \tag{2.4}$$

According to the first Eq. (2.4) and in contrary to the previous example (cf. Eq. (2.2)), the isotope enrichment factor

$$S = \frac{M^*(t) / M(t)}{M^*(0) / M(0)} = \exp\big(v(p^* - p)t\big) \; , \tag{2.5}$$

changes with time in an exponential way, thus, the isotope enrichment depends very strongly on the duration of the chemical reaction.

The depth of the chemical reaction can be expressed as a function of the concentration of isotopically non-labelled molecules or the marked molecules. If the concentration of isotopically non-labelled molecules is chosen and the quantity F is introduced as the fraction of starting molecules being converted to product, i.e.,

$$F = 1 - \frac{M(t)}{M(0)} \ ,$$
(2.6)

then *the isotope enrichment factor S* can be expressed by the following equation:

$$\ln S = \frac{1 - \alpha}{\alpha} \ln(1 - F) \ ,$$
(2.7)

where the isotope enrichment parameter α,

$$\alpha = \frac{1 - p}{1 - p^*} \ ,$$
(2.8)

does not depend on the depth of the chemical reaction. Thus, once the two macroscopic quantities F and S are determined, one can easily find the microscopic parameter $(1 - \alpha)/\alpha$ from a slope of a line $\ln(S)$ versus $\ln(1 - F)$.

Alternatively the concentration of isotopically labelled molecules M^* can be chosen to represent the degree of the chemical transformation. With the parameter

$$F^* = 1 - \frac{M^*(t)}{M^*(0)}$$
(2.9)

the isotope enrichment factor is then given by

$$\ln S = (1 - \alpha) \ln(1 - F^*) \ .$$
(2.10)

Again, one can extract the microscopic parameter α from the slope of the line $\ln(S)$ versus $\ln(1 - F^*)$.

In a similar way, one can determine a relation between the RP recombination probabilities p, p^* and observables for any other reaction scheme. Some examples which include the recombination of radicals in "cages" and in the bulk of solutions one can find in [22].

The relation between macroscopic and microscopic parameters becomes extremely simple when quantum yields of photochemical decomposition of molecules are known experimentally. The quantum yields of decomposition of the molecules M and M* (isotopically non-labelled and labelled, respectively) can correspondingly be denoted as ϕ and ϕ^*. Under the conditions specified above these quantum yields are connected with the probabilities of the in-cage RP recombination via a normalization factor, $p + \phi = 1$, $p^* + \phi^* = 1$. Then, the isotope enrichment parameter is the ratio of ϕ and ϕ^* [17],

$$\alpha = \frac{\phi}{\phi^*} . \tag{2.11}$$

From Eq. (2.7) it follows that the isotope enrichment can also be characterized by

$$\alpha_1 = \frac{1 - \alpha}{\alpha} = \frac{\phi^* - \phi}{\phi} . \tag{2.12}$$

The examples above indicate how macroscopic and microscopic parameters can be connected. For real systems the corresponding relation may become rather complicated, i.e., one has to solve appropriate kinetic equations describing a specific reaction mechanism in order to find this relation. This monograph is concerned with the physical origin of MIE. It is not our goal to present the relations connecting microscopic parameters with observables for various reaction schemes. Such formal kinetic analysis can be carried out separately for every concrete reaction mechanisms.

For the case discussed here, the main *microscopic parameter*, which directly reflects the magnetic isotope effect, is the *probability of the recombination of radical pairs*. The macroscopic quantity, which describes MIE, is the *isotope enrichment parameter*.

2.2 Radical pairs

In our previous discussion, it was assumed that the decomposition of molecules into two fragments and bimolecular processes (like the recombination of two radicals) happen through the formation of intermediate states – pairs of reactants, when these reactants are embedded in a condensed medium. This peculiarity of condensed phase reactions is of great importance for chemical kinetics. Indeed, analyzing mechanisms and kinetics of reactions, it is essential to take into account these short-lived intermediates to the same degree as free radicals, excited states, exciplexes, etc.

One may ask why radical pairs should be such universal intermediates of condensed phase reactions. How and why are they formed? Note that similar questions could be put forward concerning other reactants, not only free radicals.

There are two different reasons for pair formation.

One evident reason is related to the attractive interaction between two particles producing bound states. For example, in the case of ion-radical pairs Coulomb attraction between positively and negatively charged partners keeps them together inside the Onsager radius [5, 7]. Depending on the dielectric properties of the medium, the Onsager radius can vary from about one nanometer in water up to several tens of nanometers in non-polar systems such as paraffins. Excimers and exciplexes are another example of bound states of two partners. Of course, after their recombination two radicals also create a bound state, namely a molecule. But in this case, the binding energy is so large compared to the thermal energy that products are classified as molecules and not as radical pair states. Only the bound states that have energy level separations, comparable to thermal energies or less than thermal energies, are qualified as pair states, namely radical pairs. The possibility of forming bound pair states due to an interaction between two particles should be remembered, the specific pairs arising from Coulomb attraction, Van der Waals interaction, hydrogen bonding, weak complexing, etc., can be quite diverse.

However, there is another, rather universal, possibility for pair formation in a condensed medium which is of kinematic origin. Due to collisions with molecules of medium the free pass length of reactants is small compared to the average separation of reactants. This results in pair formation even in the absence of any attraction between the two reactants. In order to understand this kinematic effect, it is useful to start with a description of binary collisions in the condensed phase.

Suppose particles A and B are diluted in some chemically neutral solvent. Sitting on particle A, collisions with the particles B may occur as shown in Fig. 2.4. Encounters with new particles B can be described

$\longrightarrow$ Time

Fig. 2.4. One possible realization of binary collisions for a given particle A with particles B. The first vertical line in each group of encounters represents collisions with the *new partners* B while the second, third, etc. vertical line of each group depicts *re-encounters of the same two partners*. The statistics of the first collisions with the new partners B is given by the Poisson distribution, i.e., $\exp(-Kt)$. The statistics of re-encounters is described by Eq. (2.14).

by the Poisson distribution of the lifetimes between encounters. Then
the rate K of this process is given by the Smolukhovsky formula

$$K = 4\pi \, bDC_\mathrm{B} \, , \qquad\qquad (2.13)$$

where b is the encounter radius, D is the sum of diffusion coefficients
of the two reactants and C_B is the concentration of particles B.

But this Poisson distribution does not describe all binary collisions
in the condensed medium. There are additional collisions. The same two
partners can re-encounter several times before they separate. These re-
encounters are the consequence of the kinematics of diffusional motion
in the condensed phase. For the occurrence of these re-encounters two
partners do not need to have an attractive interaction. This picture of
re-encounters (see Fig. 2.4) was first simulated by E. Rabinowitch and
W. C. Wood [23] and extensively studied by R. M. Noyes [24]. During
the last twenty five years many direct and indirect proofs of re-encounters were obtained (see, e.g., reviews [5, 25]).

The statistics of re-encounters has nothing in common with the Poisson statistics of the first encounters with new partners. Instead, it be-
haves like the Green function of the diffusional motion. For example,
in the case of the continuous diffusion model re-encounters of two partners in a "cage" are described by the lifetime distribution

$$f(t) = \frac{m}{t^{3/2}} \exp\left(\frac{-\pi \, m^2}{p^2 t}\right) , \qquad\qquad (2.14)$$

where p is the probability of re-encounter and m is a parameter determined by the particle size and parameters of the diffusion motion (see,
e.g., [5, 17]). An example for this distribution is shown in Fig. 2.5.

Thus, the same two partners separate, re-approach, and re-encounter
several times. To visualize this picture, the concept of a "cage" is used
very often. Collisions with solvent molecules force two partners to separate and approach each other: solvent molecules create some sort of an
effective cage for the two reactants. This effective cage has no definite
boundaries. Instead the lifetime distribution is specified by the kinematics of the diffusion. The distribution of a lifetime between re-encounters is an exact, well determined property of two particles which enables us to consider them as well defined pair after they met each other
in a solution.

The two unpaired electrons of RPs have a total spin of either 0 or
1. The electronic states corresponding to these values of a total spin are
called the singlet and triplet states, respectively. The ability of RPs to
recombine is strongly connected with their total electron spin. The typical ground state of molecules is the singlet state. Thus, RPs will recombine preferably, if not only, from the singlet state. Recombination of trip-

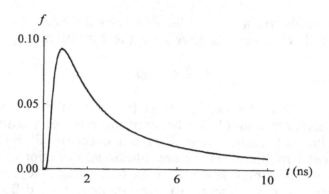

Fig. 2.5. The distribution of re-encounters, see Eq. (2.14). The parameters used for the calculations are: $p = 0.5$ and $m = 10^{-5}\, s^{1/2}$.

let RPs is far less probable. In some cases, e.g., the oxygen molecule, the ground state is the triplet state and consequently triplet RPs will recombine preferably. In the RP state, the total spin of the two unpaired electrons changes due to the spin dynamics. Spin evolution is caused by magnetic interaction of the electron spins with external magnetic fields, by hyperfine interaction of the electrons with nuclei, by Heisenberg exchange and spin-spin dipole-dipole interaction between unpaired electrons, by spin-orbit coupling, etc. (see, e.g., [5, 25, 26]). Since the spin evolution changes the relative populations of RPs in the singlet and triplet state, it modulates the ability of RPs to recombine. It is exactly this *spin evolution* which is responsible for the magnetic isotope effect in radical reactions. It will be considered in the next section.

Radical pairs might be prepared in different ways. One way is the dissociation of molecules into two radicals. In this case the two partners of a pair inherit the multiplicity of the electronic state and the mutual spatial orientation of the two radicals from the precursor molecule. As a consequence, when they come into existence, the two radicals have initially a specific alignment of unpaired electron spins. While they remain in the cage, the radical pairs remember this initial alignment of electron spins and this feature will reveal itself in the total probability of the radical pair recombination. These types of RPs are called correlated, or spin-correlated pairs, or geminate pairs. This terminology is used to indicate that two partners of the pair originate from a common precursor. In Chap. 1, several examples of spin-correlated or geminate RPs were given. Also the relations between microscopic and macroscopic parameters of MIE in Sect. 2.1 were discussed with respect to geminate pair recombination. Geminate recombination is also named cage-recombination.

Another example of RP generation is electron transfer from donor to acceptor molecule. As a result, spin-correlated ion-radical pairs are created and the above description of RP evolution applies as well.

Next we consider two radicals which encounter occasionally in solution. They will spend some time in a "cage" and thus form RPs as well. But just before the first collision within a reaction radius, the spin states of these two particles are completely uncorrelated, there is no alignment of the two unpaired electron spins. According to statistical weights, in this case RPs are formed in the singlet state with the probability 1/4 and in the triplet state with the probability 3/4. These RPs are called non-correlated or diffusion pairs. In fact, a situation with diffusion RPs is more subtle: they are non-correlated only before the first encounter of two partners. Correctly one has to say that as precursors the partners of diffusion RPs are uncorrelated. But after the very first encounter RPs, which escape a recombination, will stay some time in a "cage" and for this time they will be in spin-correlated state. Suppose RPs recombine only from the singlet state. Then encounters of partners of diffusion RPs will deplete the content of singlet RPs. As a consequence, diffusion RPs will be found preferably in the triplet state, since part of singlet RPs will be depleted by forming products of recombination. This results in a really remarkable feature of diffusion RPs. Let us imagine a homogeneous solution of radicals, which recombine from the singlet state. A quasi-stationary concentration of diffusion RPs is expected. And as a consequence of the above arguments these pairs should have polarized electron spins, because the spins of two unpaired electrons develop preferable parallel (triplet) alignment. A very interesting example of a "cage" is a micelle [20]. The volume inside the micelle can serve as cage when the lifetime of radicals is comparable or longer than the characteristic time of the diffusion across the micelle

$$\tau = \frac{L^2}{D} \, , \qquad (2.15)$$

where L is a linear dimension of the micelle, D is the mutual diffusion coefficient of radicals.

One of the most important characteristics of RPs is the statistics of re-encounters, the distribution of lifetimes between re-encounters of two partners of RPs. One example was presented above, see Eq. (2.14). Unfortunately not very much is known about the exact form of the statistics of re-encounters of partners in RPs (see, also [5]). Nevertheless one can formulate some characteristic features. In case of neutral radicals, reacting in homogeneous solutions, the asymptotics of the distribution of lifetimes between re-encounters decreases rather slowly, inversely proportional to $t^{3/2}$, see Eq. (2.14). Thus, there is a considerable part of re-encounters which happens after a long time of diffusion.

Several factors determine the distribution of re-encounters in RPs.

Kinematics of diffusion. When the length of a jump during an elementary diffusion step increases, the probability of re-encounters decreases (see, e.g., [5]). In the limit of thin gases when the length of free flight becomes much bigger than the particle size the probability of re-encounters approaches zero. Thus the kinematics of the thermal motion in thin gases does not provide background for the formation of intermediate pair states during gas phase reactions.

Interactions between partners of RPs are able to change very essentially the statistics of re-encounters. For example, Coulomb attraction between ion-radicals will raise the longtime "tale" of the distribution, i.e., re-encounters after long diffusion become more probable than for neutral reactants [7]. An attractive force between partners increases the characteristic lifetime of RPs.

The statistics of re-encounters is affected strongly by the existence of *restricted spaces for diffusion*. Restricted spaces such as the micellar interior create a sort of "super-cages" for RPs. Usually, after diffusion in restricted spaces particles will have spent more time in RP states compared to RPs in homogeneous solution. Thus systems with restricted spaces for reactants provide the possibility to increase the lifetime of RPs.

The statistics of re-encounters depends also on *chemical transformations* of radicals. Suppose that radicals can react bimolecularly with acceptors or transform monomolecularly giving adducts or new radical species, respectively. Denote the rate constant of these reactions as K. Then the distribution function of re-encounters $f(t)$ will be modified by additional factor $\exp(-Kt)$ which describes the decrease of the probability of re-encounters as a result of the chemical transformations of the radicals. Therefore these chemical transformations will decrease the in-cage lifetime of two reactants.

There is one more factor which affects the full picture of re-encounters of radicals in pairs: spin evolution of the two unpaired electrons in RPs. How the electron spin evolution contributes to the recombination of RPs will be discussed in the next sections.

The arguments outlined above give an idea about the general behaviour of re-encounters. For a quantitative discussion we need an expression of the RP lifetime, i.e., an estimate how long two partners stay in the cage and thus in the pair state.

In homogeneous solutions of neutral radicals, the characteristic lifetime of RPs is the time the two partners need to diffuse a distance comparable with their size [5]:

$$\tau = \frac{b^2}{D} , \qquad (2.16)$$

where b is the sum of the Van der Waals radii of two radicals and D is the sum of the diffusion coefficients of the two partners. In the case of oppositely charged ion-radicals this characteristic time can be estimated as the diffusion time inside the volume determined by the Onsager radius r_0 [7]

$$\tau = \frac{r_0^2}{D} \, . \tag{2.17}$$

The Onsager radius is the distance between two charged particles q_1 and q_2 for which their Coulomb energy is equal to the thermal energy

$$\frac{q_1 q_2}{r_0 \varepsilon} = kT \, . \tag{2.18}$$

Depending on the dielectric permeability ε, r_0 can vary from one nanometer to tens of nanometers. When the reactants diffuse in restricted spaces then the characteristic lifetime of RPs is determined by the size of the restricted space (see Eq. (2.15)).

As was mentioned earlier, chemical transformations of radicals reduce the lifetime in pair states. The characteristic lifetime of RPs will be in any case less than the inverse rate constant of the chemical transformations of radicals

$$\tau < \frac{1}{K} \, . \tag{2.19}$$

Depending on a number of factors such as the viscosity (or the diffusion coefficient D), on whether the radicals are neutral or carry charges, on the fact whether radicals are free to move or can only move within a restricted space, on the chemical transformations, etc., the characteristic lifetime of RPs can vary in the range 0.1–100 nanoseconds. This is the time during which spin evolution of RPs can develop and produce the magnetic isotope effect.

The discussion of the radical pair concept presented above permits us to introduce a somewhat modified phenomenological scheme for the dissociation of molecules forming free radicals or for the recombination of free radicals in a condensed phase. Usually these reactions are visualized by the scheme:

$$M \rightleftarrows \cdot R_1 + \cdot R_2 \, . \tag{2.20}$$

According to this scheme there are only two states for the free radicals R_1 and R_2: they are either separated or they are in the bound molecular state M. The essence of the previous discussion asks for the following modification of this scheme

$$M \rightleftarrows \{ \cdot R_1 \cdot R_2 \} \rightleftarrows \cdot R_1 + \cdot R_2 \, . \tag{2.21}$$

An intermediate radical pair state is added here. For reactions in the condensed phase this kind of intermediate will always be present. These pairs are very interesting creations. There is no fixed distance between the partners of the pair. There may be (but does not have to be) an attractive interaction between the two partners of the pair. What precisely specifies these intermediates is the statistics of re-encounters of the pair partners or the distribution of lifetimes between two subsequent encounters of free radicals within the recombination radius. The intermediate states offer the possibility that rather small interactions like the hyperfine interaction manifest themselves and lead to the magnetic isotope effect.

2.3 Multiplicity and reactivity of radical pairs

The phenomenological scheme (2.21) has to be further modified if the electron states are taken into account. The RPs have to be differentiated according to the multiplicity of their electronic state. The low energy level states of RPs are the singlet and triplet states. In the singlet state the total electron spin equals 0, in the triplet state it equals 1. The reason for treating S and T RPs separately is their different reactivity. Usually, RPs recombine only from the singlet state. Thus, with respect to the recombination process the RPs having different electron multiplicity possess different reactivities. On the basis of this argument the reaction mechanism (2.21) is modified as

$$M \; \rightleftharpoons \; {}^S\{ \cdot R_1 \; \cdot R_2\} \; \rightleftharpoons \; \cdot R_1 + \cdot R$$

$$\downarrow\uparrow \quad \text{(S-T transitions)}$$

$$\tag{2.22} {}^T\{ \cdot R_1 \; \cdot R_2\} \; \rightleftharpoons \; \cdot R_1 + \cdot R_2 \; .$$

From this scheme it is evident that S-T transitions are able to affect the RP's recombination since they change the population of RPs in the reactive S state.

RPs can start from the singlet state, as a result of the decomposition of singlet precursor molecule. In many photochemical reactions RPs are created from excited triplet molecules so that the initial multiplicity of RPs will also be triplet. In the case of diffusion RPs, which form during an accidental encounter of two radicals, the starting multiplicity of RPs is statistical: they form in the singlet and triplet states with the probabilities 1/4 and 3/4, respectively.

Generally, the probability of RP recombination is determined by the initial multiplicity of the RP and by the multiplicity of the recombination channel. Typically, the multiplicity of the recombination channel is the singlet.

In fact, there are interconnections between the recombination probabilities for different multiplicity of the RP precursor *provided the geminate (spin-correlated) RPs are formed within reaction radius* (see [5]). To develop these relations let us introduce some notations. ${}^{S}p_g$ and ${}^{T}p_g$ denote the geminate recombination probabilities for singlet and triplet precursors, respectively, and p_r is the probability for diffusion pair recombination. Suppose RPs can recombine only in the singlet state (note that similar arguments can be made for recombination from the triplet state, see also [5]). The evident relation here is

$$p_r = \frac{1}{4}({}^{S}p_g + 3\,{}^{T}p_g) \; . \tag{2.23}$$

However, there is also another, less obvious relation. It stems from the fact that the singlet-triplet transitions in RPs do not affect the equilibrium state. To derive this relation let us again assume that RPs recombine in the singlet state. Under this condition the ratio between the diffusion RP recombination probability and the probability of the creation of two separated radicals due to singlet molecule decomposition does not depend on the singlet-triplet evolution. The quantity $p_r/(1 - {}^{S}p_g)$ must be invariant to any changes in the singlet-triplet evolution in RPs. Suppose there are no S-T transitions in RPs. λ denotes the probability of singlet-born RP recombination. Then $p_r = \lambda/4$. The independence of the equilibrium state on S-T transitions suggests the following relation [5]

$$\frac{p_r}{1 - {}^{S}p_g} = \frac{\lambda}{4(1 - \lambda)} \; . \tag{2.24}$$

Thus, the probabilities p_r, ${}^{S}p_g$ and ${}^{T}p_g$ are not independent. Both relations (2.23) and (2.24) are valid for any isotope composition of RPs. For example, for RPs differing in their isotope composition, Eq. (2.24) can be formulated as

$$\frac{p_r}{1 - {}^{S}p_g} = \frac{p_r^*}{1 - {}^{S}p_g^*} \; , \tag{2.25}$$

where p and p^* correspond to the RPs with different isotope composition. From this relation we immediately obtain the interconnection between the isotope enrichment parameter for the singlet molecule decomposition (see Eq. (2.8))

$$\alpha_g = \frac{1 - {}^{S}p_g}{1 - {}^{S}p_g^*} \tag{2.26}$$

and the diffusion RPs recombination probabilities, since according to Eq. (2.25)

$$\alpha_{\mathrm{g}} = \frac{p_{\mathrm{r}}}{p_{\mathrm{r}}^{*}} \ . \tag{2.27}$$

If one of the RPs has no magnetic nuclei and $p_{\mathrm{r}} = \lambda/4$, then $\alpha_{\mathrm{g}} = \lambda/(4p_{\mathrm{r}}^{*})$.

The RP recombination probabilities $^{\mathrm{S}}p_{\mathrm{g}}$ and $^{\mathrm{T}}p_{\mathrm{g}}$ change with the isotope substitution in an opposite way. Thus the magnetic isotope effect will have opposite signs for singlet-born and triplet-born RPs. Indeed, let us compare the isotope substitution effect in the decomposition of the molecule from the singlet and triplet initial state. For example, di-benzyl ketone decomposes thermally from the singlet state while the photochemical decomposition happens from the triplet excited state of the molecule. Suppose that $\Delta^{\mathrm{S}}p_{\mathrm{g}}$ denotes the change of the RP recombination probability of singlet-born pairs with the isotope substitution. From Eqs. (2.23) and (2.24) we have $^{\mathrm{S}}p_{\mathrm{g}} = \lambda - 3(1 - \lambda)^{\mathrm{T}}p_{\mathrm{g}}$ or $^{\mathrm{T}}p_{\mathrm{g}} = (\lambda - {}^{\mathrm{S}}p_{\mathrm{g}})/3(1 - \lambda)$. Thus for the isotope substitution considered the change of the RP recombination probability for triplet-born pairs is $\Delta^{\mathrm{T}}p_{\mathrm{g}} = -\Delta^{\mathrm{S}}p_{\mathrm{g}}/3(1 - \lambda)$. The fact that $\Delta^{\mathrm{S}}p_{\mathrm{g}}$ and $\Delta^{\mathrm{T}}p_{\mathrm{g}}$ have opposite signs is evident: in triplet-born RPs T-S transitions increase the recombination from the singlet state, while in singlet-born pairs the same S-T transitions reduce the recombination of the singlet pairs. Diffusion RPs reveal the same sign of the isotope substitution influence as the geminate recombination of triplet-born RPs: $\Delta p_{\mathrm{r}} = (3/4)\lambda \Delta^{\mathrm{T}}p_{\mathrm{g}}$.

Thus, for isotope substitution the probability of the RP recombination changes oppositely for singlet-born to that of triplet-born RPs. Quantitatively, the absolute values of these changes are strictly interconnected as it was described above. These exact interconnections do not offer any a priori advantage for the multiplicity of RPs from the point of view of the magnetic isotope effect.

2.4 The hyperfine coupling

The magnetic isotope effect originates from the difference of the hyperfine coupling between unpaired electrons and nuclear spins after isotope substitution. There are two parts of the hf coupling: the isotropic and anisotropic interaction (see, e.g., [27, 28]).

The isotropic hf interaction is also known as contact or Fermi interaction. It is proportional to the electron spin density at the location of the nuclei. The spin Hamiltonian of the contact hfi is given by the scalar product of electron, S, and nuclear, I, spin vectors

$$\mathscr{H} = \hbar a \, \mathbf{S} \cdot \mathbf{I} \ , \tag{2.28}$$

where a is the hf coupling constant quoted in the frequency units (rad/s). From the point of view of the MIE the most important feature of the hfi is that the hf coupling constant is proportional to the magnetogyric ratio γ_n of the nucleus [27]:

$$a = \frac{8\pi}{3} g\beta \gamma_n \rho(0) , \qquad (2.29)$$

where g is the g-factor of the radical's unpaired electron, β is Bohr magneton, $\rho(0)$ is the electron spin density at the location of the nucleus. The isotropic hf coupling is an inherent property of the given radical, it does not depend on the molecular motion of the radicals as a whole. Normally there are more than one magnetic nuclei present in the free radical so the complete spin Hamiltonian of the isotropic hf coupling will contain all their contributions:

$$\mathscr{H} = \hbar \sum_k a_k \, \mathbf{S} \cdot \mathbf{I}_k . \qquad (2.30)$$

The anisotropic part of the hf coupling arises from the dipole-dipole interaction of an unpaired electron and nuclei. It has the form

$$\mathscr{H} = \hbar g\beta \gamma_n \sum_k \left(\frac{(\mathbf{S} \cdot \mathbf{I}_k)}{r_k^3} - \frac{3(\mathbf{S} \cdot \mathbf{r}_k)(\mathbf{I}_k \cdot \mathbf{r}_k)}{r_k^5} \right) , \qquad (2.31)$$

where $\mathbf{r}_k$ is the vector connecting the electron with the k-th nucleus of spin I_k. For the further discussions it is convenient to rewrite the dipole-dipole interaction Hamiltonian as (see, e.g., [29])

$$\mathscr{H} = \hbar \sum_k \frac{g\beta \gamma_n}{r_k^3} (A + B + C + D + E + F) , \qquad (2.32)$$

where

$$A = S_z I_{kz}(1 - 3\cos^2 \theta_k) ,$$

$$B = \frac{1}{2}(1 - 3\cos^2 \theta_k)(S_z I_{kz} - \mathbf{S} \cdot \mathbf{I}_k) ,$$

$$C = -\frac{3}{2}\sin \theta_k \cos \theta_k \exp(-i\varphi_k)(S_z I_{k+} + S_+ I_{kz}) ,$$

$$D = C^* = -\frac{3}{2}\sin \theta_k \cos \theta_k \exp(i\varphi_k)(S_z I_{k-} + S_- I_{kz}) ,$$

$$E = -\frac{3}{4}\sin^2 \theta_k \exp(-2i\varphi_k) S_+ I_{k+} ,$$

$$F = E^* = -\frac{3}{4}\sin^2\theta_k \exp(2i\varphi_k)\, S_- I_{k-}$$

and θ_k and φ_k are the polar and azimuth angles of the radius-vector $\mathbf{r}_k$ relative to the quantization axis which is usually determined by the direction of an external magnetic field.

Using these equations, one can estimate the value of the dipole-dipole interaction. Let us suppose that an electron interacts with a nucleus on the distance about 0.1 nm, and keep in mind that the hfi energy is quoted in the frequency units, rad/s. These units are convenient in discussing the time dependent properties. According to Eqs. (2.31) and (2.32) in the point-dipole approximation and assuming $r = 0.1$ nm one has

$$\mathcal{H}_{d\text{-}d} \approx \frac{g\beta\gamma_n}{r_k^3} = 1.76\,\gamma_n \ \ (\text{rad/s}) , \tag{2.33}$$

where γ_n is measured in units rad/(T·s). For example, Eq. (2.33) gives $\mathcal{H}_{d\text{-}d} \approx 5\cdot10^8$ rad/s for protons and ca. $8\cdot10^7$ rad/s for deuterons.

In contrast to the isotropic hf coupling the dipole-dipole interaction strongly varies with molecular motion: molecular motion causes random time modulation of the dipole-dipole interaction given by Eq. (2.32).

Information about hfi comes from magnetic resonance experiments (see, e.g., [27]). For example, the electron paramagnetic resonance spectra in liquids show splitting into several components due to isotropic hf coupling: the EPR spectra exhibit hf structure. The splitting yields the hf coupling constants. Table 2.1 lists some examples of isotropic hf coupling constants.

Table 2.1. Examples of the isotropic hf coupling constants distribution map for several free radicals.

Free radical	g-factor	Isotropic hf coupling parameters (G)	
Methyl radical CH_3	2.00255	$A_H\,(CH_3)$	$= 23.04$
Ethyl radical	2.00260	$A_H\,(CH_2)$	$= 22.38$
CH_3-CH_2		$A_H\,(CH_3)$	$= 26.87$
Allyl radical		$A_H\,(CH)$	$= 4.06$
$CH_2=CH-CH_2$		$A_H\,(CH_2)\,(1)$	$= 13.93$
		$A_H\,(CH_2)\,(2)$	$= 14.83$
NH_2	2.00481	A_H	$= 23.93$
		A_N	$= 10.33$
NO_2	1.9956	A_N	$= 47.2$

The data are taken from [30]. The coupling constants are quoted in Gauss (to convert coupling constants to the frequency units rad/s they are to be multiplied by the electron magnetogyric ratio). Note that these coupling constants will change proportionally to the nuclear magnetogyric ratios of the isotopes in isotope substitution processes.

The anisotropic part of the hfi averages to zero in non-viscous liquids, so that it does not produce any splitting in the EPR spectra. Nevertheless, also from EPR spectroscopy in liquids one can obtain information about the anisotropic part of the hfi, since dipole-dipole interaction in non-viscous liquids is responsible for paramagnetic relaxation.

The relaxation rate due to electron-nuclear dipole-dipole interaction in the absence of an external magnetic field is described by the relation [27]

$$\frac{1}{T_2} = 2W \; ,$$

$$W = \frac{2}{3} I(I+1)(g\beta\gamma_n)^2 r^{-6}\tau_0 \; , \qquad (2.34)$$

where r is the distance between the electron and the nucleus and τ_0 is the correlation time of the radical's rotational motion. Let us insert some typical values for the parameters τ_0 and r: $\tau_0 = 10^{-11}$ s and $r = 0.1$ nm. Then Eq. (2.34) gives the following estimate for the paramagnetic relaxation rate

$$\frac{1}{T_2} \approx 4\gamma_n^2\tau_0 \approx 4 \cdot 10^{-11}\gamma_n^2 \; . \qquad (2.35)$$

In this expression the nuclear magnetogyric ratio should be expressed in rad/(T·s) units. For protons Eq. (2.35) yields

$$\frac{1}{T_2} \approx 2 \cdot 10^6 \; \text{s}^{-1}. \qquad (2.36)$$

Note that the rate of the dipole-dipole interaction induced relaxation is proportional to $\gamma_n^2 I(I+1)$. This means that, e.g., in the case of H $\rightarrow$ D isotope substitution the rate of paramagnetic relaxation is scaled down by a factor of about 16.

Constants of the hfi can also be derived from EPR experiments in solids or viscous liquids, or from experiments on the Overhauser effect (see, e.g., [27, 28]).

A nucleus with spin I possesses the magnetic moment

$$\mu_n = \hbar\gamma_n\sqrt{I(I+1)} \; , \qquad (2.37)$$

where $\hbar$ is the Planck constant. The magnetic moment of a nucleus is usually quoted in units of the nuclear magneton μ_N, as μ_n/μ_N.

For isotope substitution, either the spin I of the nucleus or its magnetogyric ratio can change, or they can both change simultaneously. A list of the spin properties of some isotopes, excluding radioactive nuclei (except tritium) is presented in Table 2.2.

Table 2.2. A list of spin properties of some isotopes (data are taken from [31]).

Chemical element	Chemical symbol	Mass number	Spin	Natural abundance (% of atoms)	Magnetic moment μ_n/μ_N	Magnetogyric ratio $\gamma_n/10^7$ (rad/T·s)
Hydrogen	H	1	1/2	99.985	4.8371	26.7510
	D	2	1	0.015	1.2125	4.1064
	T*	3	1/2	–	5.1594	28.5335
Lithium	Li	6	1	7.42	1.1624	3.9366
		7	3/2	92.58	4.2035	10.396
Boron	B	10	3	19.58	2.0793	2.8748
		11	3/2	80.42	3.4702	8.5827
Carbon	C	12	0	98.892	0	0
		13	1/2	1.108	1.2162	6.7263
Nitrogen	N	14	1	99.63	0.5706	1.9324
		15	1/2	0.37	−0.4901	−2.7107
Oxygen	O	16	0	99.759	0	0
		17	5/2	0.037	−2.2398	−3.6266
		18	0	0.204	0	0
Magnesium	Mg	24	0	78.60	0	0
		25	5/2	10.11	−1.0110	−1.6370
		26	0	11.29	0	0
Silicon	Si	28	0	92.17	0	0
		29	1/2	4.71	−0.9609	−5.3141
		30	0	3.12	0	0
Sulphur	S	32	0	95.018	0	0
		33	3/2	0.750	0.8296	2.0517
		34	0	4.215	0	0
		36	0	0.017	0	0
Chlorine	Cl	35	3/2	75.53	1.0598	2.6212
		37	3/2	24.47	0.8821	2.182
Potassium	K	39	3/2	93.08	0.5047	1.2484
		41	3/2	6.91	0.2770	0.6852
Calcium	Ca	40	0	96.97	0	0
		42	0	0.64	0	0
		43	7/2	0.145	−1.4914	−1.7999
		44	0	2.06	0	0
		46	0	0.0033	0	0
		48	0	0.185	0	0
Titanium	Ti	46	0	7.99	0	0
		47	5/2	7.32	−0.9313	−1.5079
		48	0	73.99	0	0
		49	7/2	5.46	−1.2498	−1.5083
		50	0	5.25	0	0
Vanadium	V	51	7/2	99.76	5.827	7.032
Chromium	Cr	50	0	4.31	0	0
		52	0	83.76	0	0
		53	3/2	9.55	−0.6113	−1.5120
		54	0	2.38	0	0
Iron	Fe	54	0	5.84	0	0
		56	0	91.68	0	0
		57	1/2	2.17	0.1563	0.8644
		58	0	0.31	0	0
Copper	Cu	63	3/2	69.09	2.8668	7.0904
		65	3/2	30.91	3.0711	7.5958

Table 2.2 (continued).

Chemical element	Chemical symbol	Mass number	Spin	Natural abundance (% of atoms)	Magnetic moment μ_n/μ_N	Magnetogyric ratio $\gamma_n/10^7$ (rad/T·s)
Zinc	Zn	67	5/2	4.11	1.0330	1.6726
Gallium	Ga	69	3/2	60.4	2.596	6.421
		71	3/2	39.6	3.2984	8.1578
Germanium	Ge	73	9/2	7.76	−0.9693	−0.9332
Selenium	Se	77	1/2	7.58	0.9223	5.101
Bromine	Br	79	3/2	50.54	2.7089	6.7021
		81	3/2	49.46	2.9210	7.2245
Rubidium	Rb	85	5/2	72.15	1.5952	2.5829
		87	3/2	27.85	3.5391	8.7532
Strontium	Sr	87	9/2	7.02	−1.2043	−1.1594
Tin	Sn	115	1/2	0.35	−1.582	−8.7475
		117	1/2	7.61	−1.723	−9.5301
		119	1/2	8.58	−1.8029	−9.9707

The magnetogyric ratio of different nuclei shows variation by nearly two orders of magnitude: for example, the magnetogyric ratio of ^{3}H (or T) is 62.273 times bigger than the magnetogyric ratio of ^{197}Au. The magnetic moments change with isotope substitution as well. For example, for the H → D isotope substitution the magnetogyric ratio decreases 6.5145 times while the nuclear magnetic moment decreases 3.9894 times (see Table 2.2). For the substitution ^{12}C → ^{13}C the nuclear magnetic moment changes from zero to a finite quantity (see Table 2.2), etc.

Concluding this brief outline, the hyperfine interaction in free radicals is characterized by coupling constants in the range of 10^7–10^{10} rad/s. It is proportional to the magnetogyric ratio of nuclei, and therefore changes accordingly with isotope substitution.

2.5 Singlet-triplet transitions in radical pairs induced by the hyperfine interaction

The magnetic isotope effect in radical reactions originates from the nonradiative intersystem S-T transitions in radical pairs induced by the hfi. Therefore it seems reasonable to present some examples of the S-T transition dynamics for model systems. First, we need appropriate basis functions. For RP recombination it is convenient to use the singlet and triplet states of two electron spins as a basis:

$$S = \frac{1}{\sqrt{2}}\left(\left|\frac{1}{2}, -\frac{1}{2}\right\rangle - \left|-\frac{1}{2}, \frac{1}{2}\right\rangle\right),$$

$$T_{+1} = \left|\frac{1}{2}, \frac{1}{2}\right\rangle ,$$

$$T_0 = \frac{1}{\sqrt{2}}\left(\left|\frac{1}{2}, -\frac{1}{2}\right\rangle + \left|-\frac{1}{2}, \frac{1}{2}\right\rangle\right) ,$$

$$T_{-1} = \left|-\frac{1}{2}, -\frac{1}{2}\right\rangle , \tag{2.38}$$

where $|1/2\rangle$ and $|-1/2\rangle$ are the eigenfunctions of the operators S_z of each of the electrons corresponding to the eigenvalues $1/2$ and $-1/2$, respectively. When nuclear spins are included, the basis functions will be chosen as the external product of the functions (2.38) and the functions of nuclear spins describing a set of projections of nuclear spins onto the quantization axis $|m_1, m_2, m_3, ...\rangle$. For example, for RPs with one magnetic nucleus the basis functions chosen are

$$|S, m\rangle, \ |T_{+1}, m\rangle, \ |T_0, m\rangle, \ |T_{-1}, m\rangle , \tag{2.39}$$

where $m = I, I-1, ..., -I$ for a nucleus of spin I.

In radical pairs, the S and T states are degenerate or quasi-degenerate (see, e.g., Fig. 1.1). When the separation between radicals is larger than about one nanometer, the spin-spin dipole-dipole and Heisenberg exchange interactions between unpaired electrons are small compared to the hfi in free radicals. In non-viscous solvents the anisotropic hfi is averaged out to zero due to rotational diffusion of radicals. Its effect on S-T transitions of RPs may also be neglected. Under these conditions, the spin Hamiltonian of RPs will only include the isotropic hf coupling and the interaction of electron spins with external magnetic fields. In the presence of the external magnetic field B_0 it can be written as (note that here hfi constants a are expressed in frequency units rad/s)

$$\mathcal{H} = \hbar\left(\omega_A S_{Az} + \omega_B S_{Bz} + \sum_k a_k \mathbf{S}_A \cdot \mathbf{I}_k + \sum_n a_n \mathbf{S}_B \cdot \mathbf{I}_n\right) , \tag{2.40}$$

where the Larmor frequencies of the unpaired electrons are expressed via their g-factors as

$$\omega_A = \frac{g_A \beta}{\hbar} B_0 ,$$

$$\omega_B = \frac{g_B \beta}{\hbar} B_0 . \tag{2.41}$$

Suppose there is only one nucleus of spin $I = 1/2$ and $g_A = g_B$. In this simple model system the transition matrix elements of the Hamiltonian, mixing the S and T states of RPs, are:

$$\left\langle T_0, \frac{1}{2} \left| \mathcal{H} \right| S, \frac{1}{2} \right\rangle = \hbar \frac{a}{4} \; ,$$

$$\left\langle T_0, -\frac{1}{2} \left| \mathcal{H} \right| S, -\frac{1}{2} \right\rangle = -\hbar \frac{a}{4} \; ,$$

$$\left\langle T_{+1}, -\frac{1}{2} \left| \mathcal{H} \right| S, \frac{1}{2} \right\rangle = -\hbar \frac{a}{2\sqrt{2}} \; ,$$

$$\left\langle T_{-1}, \frac{1}{2} \left| \mathcal{H} \right| S, -\frac{1}{2} \right\rangle = \hbar \frac{a}{2\sqrt{2}} \; . \qquad (2.42)$$

These equations explicitly state that, in principle, the isotropic hfi is able to mix the two singlet $|S, 1/2\rangle$ and $|S, -1/2\rangle$ states with triplet states. However, it should be noted that two of the triplet states, namely $|T_{+1}, 1/2\rangle$ and $|T_{-1}, -1/2\rangle$, are not involved in the S-T mixing process caused by the isotropic hfi. As one can see from Eqs. (2.42) S-T_0 transitions do not change the projection of the nuclear spin, while S-T_{+1} or S-T_{-1} transitions are accompanied by a flip or flop of the nuclear spin. Suppose the model system considered starts from the singlet state. The time evolution of the spin density matrix is given by the equation [32]:

$$\frac{\partial \rho}{\partial t} = -\frac{i}{\hbar} [\mathcal{H}, \rho] \; . \qquad (2.43)$$

The solution of this equation for the model under discussion leads to the following results: The subensemble of the RPs with $+1/2$ projection of the nuclear spin converts to the triplet states $|T_0, 1/2\rangle$ and $|T_{+1}, -1/2\rangle$ as

$$n\left(T_0, \frac{1}{2} \right) \equiv p\left(\left| S, \frac{1}{2} \right\rangle \rightarrow \left| T_0, \frac{1}{2} \right\rangle \right)$$

$$= \frac{1}{4} \left(1 + \cos^2 \frac{Wt}{2} + \frac{\omega^2}{W^2} \sin^2 \frac{Wt}{2} \right.$$

$$\left. - 2 \cos \frac{Wt}{2} \cos \frac{(\omega - a)t}{2} - 2 \frac{\omega}{W} \sin \frac{Wt}{2} \sin \frac{(\omega - a)t}{2} \right) ,$$

$$n\left(T_{+1}, -\frac{1}{2} \right) \equiv p\left(\left| S, \frac{1}{2} \right\rangle \rightarrow \left| T_{+1}, -\frac{1}{2} \right\rangle \right) = \frac{1}{2} \frac{a^2}{W^2} \sin^2 \frac{Wt}{2} \; , \qquad (2.44)$$

where $W^2 = a^2 + \omega^2$, and $\omega \equiv \omega_A = \omega_B$.

The subensemble of the RPs with $-1/2$ projection of the nuclear spin converts to the triplet states $|T_0, -1/2\rangle$ and $|T_{-1}, 1/2\rangle$ as

$$
n\left(T_0, -\frac{1}{2}\right) \equiv p\left(\left|S, -\frac{1}{2}\right\rangle \to \left|T_0, -\frac{1}{2}\right\rangle\right)
$$

$$
= \frac{1}{4}\left(1 + \cos^2\frac{Wt}{2} + \frac{\omega^2}{W^2}\sin^2\frac{Wt}{2}\right.
$$

$$
\left. - 2\cos\frac{Wt}{2}\cos\frac{(\omega+a)t}{2} - 2\frac{\omega}{W}\sin\frac{Wt}{2}\sin\frac{(\omega+a)t}{2}\right),
$$

$$
n\left(T_{-1}, \frac{1}{2}\right) \equiv p\left(\left|S, -\frac{1}{2}\right\rangle \to \left|T_{-1}, \frac{1}{2}\right\rangle\right) = \frac{1}{2}\frac{a^2}{W^2}\sin^2\frac{Wt}{2}. \tag{2.45}
$$

These solutions allow us to point out several observations. When the intensity of the external field increases and the Larmor frequencies of the electrons become large compared to the hf coupling constant, i.e., $\omega > a$, the populations of $|T_{+1}, -1/2\rangle$ and $|T_{-1}, 1/2\rangle$ states diminish as $(a/\omega)^2 < 1$. This result is expected since at high external magnetic fields the T_{+1} and T_{-1} states are rather far removed from the degenerated S and T_0 states of the RPs due to the Zeeman splitting of the triplet sublevels. In this case the hfi is not able to transform RPs from the singlet state to T_{+1} and T_{-1} triplet states. Thus, at high magnetic field the singlet RPs transform only to the T_0 state, in fact, the RPs oscillate between S and T_0 state with a frequency determined by the isotropic hyperfine coupling constant. For $\omega > a$ we obtain from Eqs. (2.44) and (2.45)

$$
n(T_0) \equiv p(|S\rangle \to |T_0\rangle) = \sin^2\frac{at}{4}. \tag{2.46}
$$

For low magnetic field all triplet sublevels are close to the singlet state so that hfi can considerably mix the singlet state with all triplet states. These statements are visualized in Fig. 2.6.

Singlet-triplet transitions given by Eqs. (2.44) and (2.45) are simplified for zero external field. Note that the Earth's magnetic field of about $0.5 \cdot 10^{-4}$ T may be considered as zero field when discussing MIE. At $\omega = 0$ Eqs. (2.39) and (2.40) give equal populations of the triplet sublevels for singlet-born RPs:

$$
n(T_0) \equiv p(|S\rangle \to |T_0\rangle) = \frac{1}{4}\sin^2\frac{at}{2},
$$

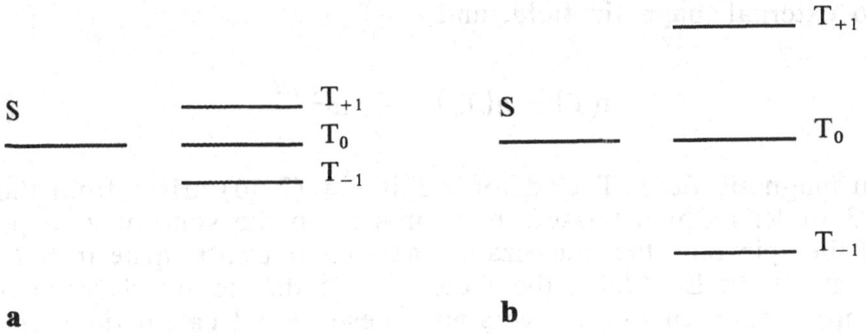

Fig. 2.6. Scheme of the RP's energy levels at low (**a**) and high (**b**) magnetic field.

$$n(T_{+1}) = p(|S\rangle \to |T_{+1}\rangle) = \frac{1}{4}\sin^2\frac{at}{2} \ ,$$

$$n(T_{-1}) = p(|S\rangle \to |T_{-1}\rangle) = \frac{1}{4}\sin^2\frac{at}{2} \ . \tag{2.47}$$

The total population of triplet states is equal to

$$n(T) = p(|S\rangle \to |T\rangle) = \frac{3}{4}\sin^2\frac{at}{2} \ . \tag{2.48}$$

Similar to the high field case (see Eq. (2.41)) RPs oscillate between the singlet and triplet state. However, the frequency of the S-T dynamics in the low field case is two times higher than in the high field case (compare Eqs. (2.46) and (2.48)). The remarkable feature of the low field S-T transitions is that, according to Eq. (2.48), RPs never totally convert to the triplet state, $n(T) \le 3/4$. In low external fields three channels of S-T transitions (S $\leftrightarrow$ T_{+1}, T_0, T_{-1}) operate in contrast to the high field limit where only one (S $\leftrightarrow$ T_0) channel acts efficiently. This single channel provides the total conversion of the singlet RPs to the triplet state in the high field case while the three channels in the low field situation are not able to convert all singlet-born RPs to the triplet state. This is due to interference between the different channels, which is connected with the RP spin conservation rules.

Analogously, the singlet-triplet transition dynamics of any RPs can be derived using the spin Hamiltonian (see, e.g., Eq. (2.40)) and solving the equation of motion for the density matrix (2.43). For example, the solution for singlet-born RPs containing one nucleus with the spin $I = 1$ gives

$$n(T) = \frac{8}{9}\sin^2\frac{3at}{4} \tag{2.49}$$

at zero external magnetic field, and

$$n(\text{T}) = n(\text{T}_0) = \frac{2}{3}\sin^2\frac{at}{2} \tag{2.50}$$

at high magnetic field. The factor 2/3 in Eq. (2.50) arises from the fact that 1/3 of RPs which possess nuclear spin in the state of zero projection of the spin onto the quantization axis cannot participate in S-T transitions at all. In Eq. (2.49) the factor 8/9 is due to interference of the different channels of S-T conversion. These model calculations are useful to get a better impression about the effect of an isotope substitution on the singlet-triplet dynamics. Let us consider the recombination of a geminate singlet-born RP which has only one magnetic nucleus, e.g., a proton. The probability to find RPs in the singlet (reactive!) state at zero magnetic field is given by (see Eq. (2.48))

$$n(\text{S}) = 1 - n(\text{T}) = 1 - \frac{3}{4}\sin^2\frac{a_{\text{H}}t}{2} \,. \tag{2.51}$$

The H → D isotope substitution changes the population of the RPs in the singlet state to (see Eq. (2.49))

$$n(\text{S}) = 1 - n(\text{T}) = 1 - \frac{8}{9}\sin^2\frac{3a_{\text{D}}t}{4} \,. \tag{2.52}$$

The hyperfine coupling constants for H and D isotopes change in proportion to their magnetogyric ratios (see Table 2.2). Suppose, for instance, that the hfi constant in the case of the proton containing radical pair is equal to 1 mT or $1.76 \cdot 10^8$ rad/s in frequency units. Then the hfi needs about 20 nanoseconds to convert this RP from the singlet state to the triplet state (see Fig. 2.7a). After H → D isotope substitution the hfi constant reduces about 6.51 times and the value of the nuclear spin increases, thus the hfi requires about 80 nanoseconds for S-T conversion of the deuterated RP (see Fig. 2.7b).

If the free radicals have many non-equivalent magnetic nuclei the contribution of the hfi to the RP's spin dynamics can be calculated in the so-called semiclassical approximation [33] where classical vectors are substituted for nuclear spin moment operators in the hfi spin Hamiltonian. Suppose there is an RP in which one of the partners has no magnetic nuclei while the other partner has many magnetic nuclei. In this case the singlet state population of *triplet-born* RPs at zero external magnetic field is expressed as [33]

$$n(\text{S}) = p(\text{T} \to \text{S}) = \frac{1}{6}(1 - c(t)) \,,$$

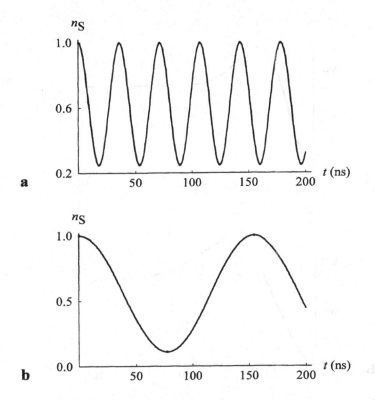

Fig. 2.7. Time evolution of the singlet state population of singlet-born RPs coupled to one magnetic nucleus at zero magnetic field. **a** Curve corresponding to the H-containing RPs, **b** result for the D-containing RPs. The hf coupling parameters are: $a_H = 1.76 \cdot 10^8$ rad/s and $a_D = 0.270 \cdot 10^8$ rad/s.

$$c(t) = \left(1 - (\gamma_e A_{ef} t)^2\right) \exp\left(-\frac{(\gamma_e A_{ef} t)^2}{2}\right) , \qquad (2.53)$$

where γ_e is the electron magnetogyric ratio, $\gamma_e = 1.76 \cdot 10^{11}$ rad/(T·s), and A_{ef} is the effective hf coupling constant quoted in Tesla

$$A_{ef}^2 = \frac{1}{3}\sum_k A_k^2 I_k(I_k + 1) . \qquad (2.54)$$

An example of the time dependence of $n(S)$ according to (Eq. (2.53)) is shown in Fig. 2.8.

Figure 2.8 demonstrates that for H → D isotope substitution the rate of triplet-singlet conversion decreases four times corresponding to the decrease of the nuclear magnetic moment.

Consider one more example of S-T evolution for RPs in zero magnetic field. Let us assume that one of the pair partners possesses many

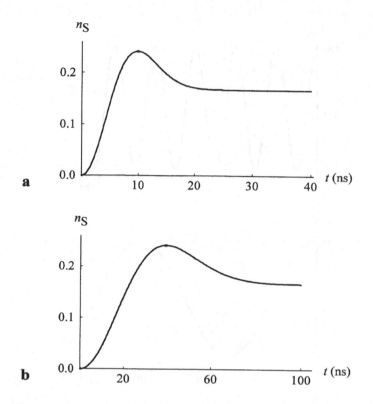

Fig. 2.8. Time dependence of the singlet state population for triplet-born RPs. **a** Curve corresponding to the totally protonated radical with $A_{ef} = 1$ mT (or $a_{ef} = 1.76 \cdot 10^8$ rad/s). **b** Time dependence of singlet population for the same RPs after H → D isotope substitution. Note that the curves are presented in different time scales.

magnetically non-equivalent nuclei while the other partner has only one magnetic nucleus with spin $I = 1/2$. For this model system the probability to find triplet-born RPs in the singlet state is equal to

$$n(\text{S}) = p(\text{T} \to \text{S}) = \frac{1}{4} \sin^2 \frac{\gamma_e A t}{2} + \frac{1}{6} (1 - c(t)) \cos^2 \frac{\gamma_e A t}{2}, \quad (2.55)$$

where as before the isotropic hf coupling constant A of the one nucleus with spin $I = 1/2$ is quoted in magnetic field intensity units (Tesla), $c(t)$ is given by Eq. (2.53).

For example, this model allows us to compare T-S dynamics in RPs which form during the photochemical decomposition of the $^{12}\text{C} \to {}^{13}\text{C}$ substituted dibenzyl ketone molecules (see Fig. 2.1). In the case of the decomposition of DBK molecules containing only ^{12}C isotopes, T-S transitions in RPs are induced by the hf coupling with all protons of the free radical CH_2Ph (see Fig. 2.1) and they are described by Eq. (2.53), the effective coupling constant being equal to $A_{ef} = 1.23$ mT. The carbon isotope substitution adds the hf coupling with the ^{13}C nucleus. The coupling with the carbon nucleus is strong when the substitution takes place in the CO or CH_2 groups of DBK. For the RPs $\{PhCH_2{}^{13}CO \cdot \cdot CH_2Ph\}$ with the magnetic nucleus ^{13}C in the CO group the hf coupling constant is equal to $A_C \cong 12.5$ mT, while for the RPs $\{Ph^{13}CH_2CO \cdot$

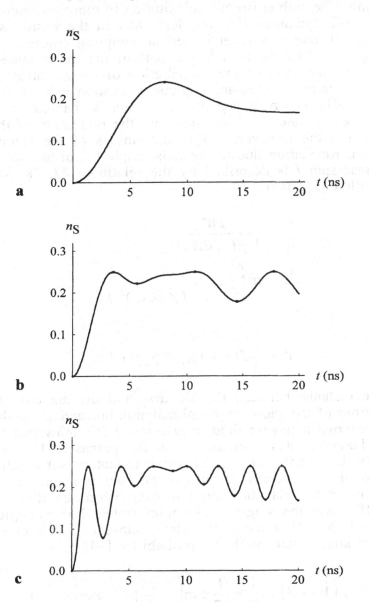

Fig. 2.9. Time evolution of the singlet state population and its variation with isotope substitution. The external magnetic field is assumed to be zero. In fact the result will be practically the same in the Earth's field. **a** Curve corresponding to $\{PhCH_2CO \cdot \cdot CH_2Ph\}$ pairs, **b** curve corresponding to $\{Ph^{13}CH_2CO \cdot \cdot CH_2Ph\}$ pairs, and **c** curve corresponding to $\{PhCH_2{}^{13}CO \cdot \cdot CH_2Ph\}$ pairs.

$\cdot CH_2Ph\}$ which contain ^{13}C in the CH_2 group $A_C \cong 5$ mT. For these two RPs the T-S dynamics can be calculated using Eq. (2.55). The results are shown in Fig. 2.9.

The results presented in Fig. 2.9 indicate clearly that the insertion of the magnetic carbon isotope increases remarkably the rate of triplet-singlet mixing. The carbon isotope substitution in other positions in DBK changes the S-T dynamics also but less than in the positions considered for Fig. 2.9, due to smaller hyperfine coupling constants.

For long-lived RPs the anisotropic part of the hfi becomes important. In non-viscous solvents the contribution of the anisotropic hfi can be described in terms of the paramagnetic relaxation induced by it (see, e.g., [27, 28, 34]). The paramagnetic relaxation is characterized by two relaxation times: T_1 and T_2. They describe the relaxation of the longitudinal (T_1) and the transverse (T_2) components of the electron spin magnetization. Relaxation due to the anisotropic hfi of an electron spin with a nuclear spin I is described by the relations [27, 28, 34] (compare also with Eq. (2.34))

$$\frac{1}{T_1} = \frac{2W}{1 + (\gamma_e B_0 \tau_0)^2} ,$$

$$\frac{1}{T_2} = W \left(1 + \frac{1}{1 + (\gamma_e B_0 \tau_0)^2} \right) , \qquad (2.56)$$

where

$$W = \frac{2}{3} I(I + 1)(g\beta_e \gamma_n)^2 r^{-6} \tau_0$$

and r is the distance between the electron and the nucleus, τ_0 is the correlation time of the radical's rotational motion, and B_0 is the intensity of the external magnetic field. Equations (2.56) show that the paramagnetic relaxation rates decrease when B_0 increases. The relaxation rates given by Eqs. (2.56) are very sensitive to isotope substitution. They are proportional to the square of the isotope nuclear magnetic moments.

The paramagnetic relaxation tends to establish an equilibrium distribution of RPs over the singlet and triplet states. For example, if the initial state of the RP is the triplet state, subsequently it may be observed in the singlet state with the probability [34]

$$n(S) = p(T \to S) = \frac{1}{4} - \frac{1}{12} \exp\left(-\frac{t}{T_1}\right) - \frac{1}{6} \exp\left(-\frac{t}{T_2}\right) , \qquad (2.57)$$

where $1/T_1 = 1/T_{1A} + 1/T_{1B}$ and $1/T_2 = 1/T_{2A} + 1/T_{2B}$ are the sums of the relaxation rates of the individual RP partners A and B. One can see that at $t > T_1, T_2$, $n(S) \to 1/4$ which is the statistical portion of RPs in singlet state in the high temperature limit considered.

Singlet-triplet evolution of RPs induced by the hyperfine coupling in a number of model systems may be found in [5]. All calculations show that in the RP state the isotropic hyperfine interaction mixes the singlet and triplet states on a time scale determined by the hf coupling constants. Another property of the hfi induced S-T transitions which should be emphasized is that the efficiency of these transitions depends on the strength of the external magnetic field, a large Zeeman splitting of the triplet substates switches off the S-T_{+1} and S-T_{-1} channels. The contribution of the anisotropic part of the hfi to the singlet-triplet evolution is also very sensitive to isotope composition of radicals and depends on the intensity of the external field. It is expected that the anisotropic hfi will be of significance either in viscous systems (when the rotational motion of radicals does not average this interaction efficiently to zero) or for long-lived RPs like RPs in micelle (when even rather slow relaxation processes will have enough time to act).

Depending on the starting multiplicity of RPs one has to consider either the S $\to$ T (for singlet-born RPs) or T $\to$ S (for triplet-born RPs) transitions. However the efficiencies of these processes are not independent:

$$p(T \to S) = \frac{1}{3} p(S \to T) . \tag{2.58}$$

The factor 1/3 arises from the 1/3 population of the triplet sublevels of triplet-born RPs.

2.6 Manifestations of radical pair spin conservation rules in singlet-triplet mixing induced by the isotropic hyperfine coupling

The previous discussion of the isotropic hfi induced S-T transitions has shown that the conversion is sometimes not complete (see, e.g., Eqs. (2.48) and (2.49)). Now we discuss how this behaviour is connected with the conservation of the total spin of RPs and of its projection onto the quantization axis [35].

Consider geminate RP recombination. Assume that RP starts from triplet precursor and can recombine only via its singlet state. We also assume that the isotropic hf coupling is the basic mechanism of RP S-T mixing. This assumption is well confirmed in low magnetic field for the great majority of organic free radicals. This turns out to be also the most favourable for the magnetic isotope effect. Indeed, let us compare two RPs differing in their isotope composition. In one of them the radicals are assumed to have no magnetic nuclei, the hfi is zero and the RP will remain in the T state, $^T p_g = 0$. In the other RP, after the isotope substitution, e.g., ^{13}C for ^{12}C, the interaction of the unpaired electrons with the magnetic nuclei induces transitions of the RP to its reactive S state. As a result, the RP recombines with the probability $^T p_g \neq 0$. The spin Hamiltonian, including isotropic hf coupling, the exchange interaction and Zeeman interactions of the RP's unpaired electrons is (see Eq. (2.40))

$$\mathcal{H} = \hbar \left\{ (\omega_A S_{Az} + \omega_B S_{Bz}) + \sum_k a_{Ak} \mathbf{S}_A \cdot \mathbf{I}_k \right.$$
$$\left. + \sum_n a_{Bn} \mathbf{S}_B \cdot \mathbf{I}_n - J(r) \left(\frac{1}{2} + 2 \mathbf{S}_A \cdot \mathbf{S}_B \right) \right\}, \qquad (2.59)$$

where $J(r)$ is the exchange integral depending on the distance r between the two radicals. Let N be the number of all possible states of the RP nuclear spins,

$$N = \prod_k \prod_n (2I_k + 1)(2I_n + 1) . \qquad (2.60)$$

In the case considered, all sublevels of the triplet RPs are populated initially with the same probability $1/3N$. The Hamiltonian (2.59) induces $T \rightarrow S$ mixing in the system. Let L be the number of triplet sublevels connected with the singlet state by non-zero transition matrix elements $\mathcal{H}_{TS}$ (see, e.g., Eqs. (2.42)). In this case the maximum probability of $T \rightarrow S$ mixing is

$$p(T \rightarrow S)_{max} = \frac{L}{3N} . \qquad (2.61)$$

It can be less than $L/3N$ due to interference of the L triplet-singlet transition channels.

The relation (2.61) sets the general limit that can be reached at a given isotope composition. The quantity L can be found using the following arguments.

Let us start from the case of zero external magnetic field. This case is of particular interest since the Earth's magnetic field can be approximated as zero from the viewpoint of the magnetic isotope effect. In this case the total moment of the RP spins:

$$\Sigma = \mathbf{S}_A + \mathbf{S}_B + \sum_k^{(A)} \mathbf{I}_k + \sum_n^{(B)} \mathbf{I}_n , \qquad (2.62)$$

commutes with the Hamiltonian (2.59)

$$[\mathcal{H}, \Sigma^2] = 0 \ . \tag{2.63}$$

Equation (2.63) means that the total moment of the RP spins is conserved during S-T transitions. Not all triplet RP states are mixed with the singlet one: only the transitions with conservation of Σ^2 can occur and hence $L < 3N$. To find the quantity L let us sum up the moments of all RP nuclei. Let the total nuclear spin have the values $J_1, J_2, ..., J_n$. The sum of the unpaired RP electron spins gives $S = 0$ or $S = 1$. The sum of all RP spin moments allows the states with the total spin $\Sigma = J_1, J_2, ..., J_n$ when unpaired electrons are in the singlet state ψ_S, and the states with $\Sigma = J_1 + 1, J_1, |J_1 - 1|, J_2 + 1, J_2, ..., |J_n - 1|$ when unpaired electrons are in the triplet state ψ_T. Due to spin conservation only those ψ_T states contribute to S-T transitions for which Σ coincides with one of the Σ values for ψ_S. The number of states selected in this way is L. This procedure is illustrated now for some examples. For an RP with one magnetic nucleus of spin I the summation results in states with a total spin $\Sigma = I$ and $\Sigma = I + 1$, $\Sigma = |I - 1|$. The total spin $\Sigma = I$ is possible with states having $S = 0$ and $S = 1$. Due to the spin conservation, see Eq. (2.63), RPs with $\Sigma = I + 1, I - 1$ will remain in the initial triplet (*non reactive*) state. T $\rightarrow$ S transitions can only occur from $(2I + 1)$ triplet sublevels with $\Sigma = I$. As a result, $L = N = 2I + 1$. Hence, for an RP with one magnetic nucleus

$$p(T \rightarrow S)_{max} = 1/3 \ , \tag{2.64}$$

irrespective of the nuclear spin value, $I \geq 1/2$.

The relation (2.64) can be employed to obtain $p(T \rightarrow S)_{max}$ for an RP in which one of the radicals has an arbitrary number of magnetically equivalent nuclei [35]. For an RP with an odd number of nuclei having a half-integer spin it follows $p(T \rightarrow S)_{max} = 1/3$ and for an RP with an even number of nuclei having a half-integer spin, or for an RP with any $(n > 1)$ number of nuclei having an integer spin, the total nuclear spin may be zero and hence $p(T \rightarrow S)_{max} < 1/3$. In systems with non-equivalent nuclei the probability of the triplet-singlet transitions increases. According to [35] one obtains

$$p(T \rightarrow S)_{max} = 1 - \frac{2nI + 3}{3(2I + 1)^n} \tag{2.65}$$

for an RP with $n \geq 2$ magnetically non-equivalent nuclei of equal spin I. Using this relation one can estimate how the probability of S-T transitions changes with isotope substitution. For example, in the case of H $\rightarrow$ D isotope substitution $p(T \rightarrow S)_{max}$ changes from

$$p(T \to S; \frac{1}{2})_{max} = 1 - \frac{n+3}{3 \cdot 2^n} \qquad (2.66)$$

to

$$p(T \to S; 1)_{max} = 1 - \frac{2n+3}{3^{n+1}} . \qquad (2.67)$$

Examples for different numbers n of equivalent nuclei are given in Table 2.3.

The maximum probabilities for deuterated RPs turn out to be bigger than for protonated RPs. This reflects the increase in number of the S-T mixing channels with increasing the nuclear spin value. Note, however, that for H → D isotope substitution, the magnetic moment of the nuclei decreases about four times. As a consequence, the rate of S-T transitions decreases. Deuterated RPs need more time to reveal the limits given in Table 2.3. Thus, two competitive tendencies are associated with H → D isotope substitution: an increase in the fraction of RP subensembles in which S-T transitions are allowed and, due to the reduction of the isotropic hfi constants by a factor of 6.5, a decrease in the rate of reaching the maximum possible value of the T-S transition probability.

In fields $B_0 \neq 0$, the total electron and nuclear spin is not conserved, $[\mathscr{H}, \Sigma^2] \neq 0$. However, the projection of the total spin onto the external field direction termed z commutes with the spin Hamiltonian $\mathscr{H}$ and thus S-T transitions can only occur with Σ_z conserved. This imposes a reduced limitation on S-T transitions as compared to the conservation of Σ^2 in zero field. As a result, $p(T \to S)_{max}$ must increase with B_0. Let us consider, for example, the simplest case of an RP with one nucleus of spin $I = 1/2$. The total electron and nuclear spin has the projections $\Sigma_z = \pm 1/2$ in the singlet electron state and $\Sigma_z = \pm 1/2$, $\pm 1/2$, $\pm 3/2$ in the triplet electron state. Thus four of the six triplet sublevels can participate in S-T transitions, while transitions are forbidden for the $\Sigma_z = \pm 3/2$ states. As a result, $p(T \to S)_{max}$ can increase up to 2/3 in a magnetic field, i.e., twice as high as in zero field. For an

Table 2.3. The maximum possible probabilities of the T → S transitions calculated from Eqs. (2.66) and (2.67) [35].

n	$I = 1/2$, Eq. (2.66)	$I = 1$, Eq. (2.67)
1	0.33	0.33
2	0.58	0.74
3	0.75	0.89
4	0.85	0.95
5	0.92	0.98
10	0.995	0.9999

RP with one nucleus of an arbitrary spin I in intermediate fields $B_0 \cong$ $\cong A_{ef}$ one obtains

$$p(T \to S)_{max} = 1 - \frac{2}{3(2I+1)} . \tag{2.68}$$

Similar considerations show that for any RP there are only two triplet sublevels with $\Sigma_z = \pm(1 + \Sigma_k I_k)$ for which the transitions to the S state are forbidden due to the requirement of Σ_z conservation. Therefore, e.g., in a case of a system with n equivalent nuclei,

$$p(T \to S)_{max} = 1 - \frac{2}{3(2I+1)^n} . \tag{2.69}$$

For RPs with n nuclei having a spin $I = 1/2$ and m nuclei with spin $I = 1$

$$p(T \to S)_{max} = 1 - \frac{1}{2^{n-1} \cdot 3^{m+1}} . \tag{2.70}$$

As an example, the $p(T \to S)_{max}$ values for RPs with n equivalent nuclei with $I = 1/2$ and $I = 1$ are given in Table 2.4.

Comparison of Tables 2.3 and 2.4 shows that $p(T \to S)_{max}$ reaches higher values in intermediate magnetic fields than in zero field.

In high fields, when $B_0 \gg A_{ef}$, the triplet T_{+1} and T_{-1} sublevels are out of resonance with the S state. Only T_0-S transitions are allowed (see Fig. 2.6). As a result, only one-third of the triplet pairs participate in T-S transitions and

$$p(T \to S)_{max} \leq 1/3 . \tag{2.71}$$

$p(T \to S)_{max}$ will be below 1/3, if the projection of the total local hfi field onto the external magnetic field direction is zero for certain nuclear spin orientations. For instance, in the case of an RP with one

Table 2.4. $p(T \to S)_{max}$ values for radical pairs with n equivalent nuclei with $I = 1/2$ (in column 2) and $I = 1$ (in column 3) calculated using Eq. (2.69) [35].

n	$p(T \to S; 1/2)_{max}$	$p(T \to S; 1)_{max}$
1	0.667	0.778
2	0.833	0.926
3	0.917	0.975
6	0.989	0.992
8	0.997	0.997

nucleus of integer spin, T_0-S transition is not possible for the RP sub-ensemble with a zero projection of the nuclear spin so that

$$p(T \to S)_{max} = \frac{2I}{3(2I+1)} . \tag{2.72}$$

For an RP with an odd number of magnetically equivalent nuclei having a half-integer spin

$$p(T \to S)_{max} = 1/3 . \tag{2.73}$$

In the case of an even number of nuclei of spin $I = 1/2$,

$$p(T \to S)_{max} = \frac{1}{3}\left(1 - \frac{C_n^{n/2}}{2^n}\right) , \tag{2.74}$$

where the number of combinations $C_n^{n/2}$ gives the number of states wherein the local hfi fields of various nuclei compensate each other to zero. According to Eq. (2.74) $p(T \to S)_{max} \to 1/3$ with increasing n, see Table 2.5.

For RPs with non-equivalent nuclei the local hfi fields from various nuclei can only accidentally compensate each other to zero. Hence, for such systems in high magnetic fields one should expect

$$p(T \to S)_{max} = 1/3 . \tag{2.75}$$

The above results give an idea of the tendency of S-T mixing variation with isotope substitution when the isotropic hfi is the basic mechanism of RP singlet-triplet mixing. They also demonstrate the field dependence of the singlet-triplet transitions in RPs. The results obtained for some simple systems with n non-equivalent nuclei of spin I are listed in Table 2.6.

From Table 2.6, we can infer that in the case of RPs with a few magnetically non-equivalent nuclei, $p(T \to S)_{max}$ *passes through a maximum* with increasing external magnetic field. For RPs with a sufficiently large number of about 10 or more magnetically non-equivalent nuclei

Table 2.5. $p(T \to S)_{max}$ values for high external magnetic field calculated from Eq. (2.74) [35].

	n						
	2	4	6	8	10	12	20
$p(T \to S)_{max}$	0.167	0.208	0.229	0.242	0.251	0.258	0.275

Table 2.6. Comparison of $p(T \to S)_{max}$ values calculated at zero, intermediate and high external magnetic fields for RPs with non-equivalent nuclei [35].

n	I	$B_0 = 0$	$B_0 \cong A_{ef}$	$B_0 \gg A_{ef}$
1	1/2	0.33	0.67	0.33
1	1	0.33	0.78	0.22
2	1/2	0.58	0.83	0.17
2	1	0.74	0.93	0.22
3	1/2	0.75	0.92	0.33
3	1	0.89	0.97	0.25

$p(T \to S)_{max}$ approaches 1 at zero and intermediate external magnetic fields and will reduce from 1 to 1/3 when the field increases from zero to high values.

It is this field dependence of singlet-triplet transitions in RPs which is responsible for the field dependence of the magnetic isotope effect. To illustrate this statement, let us consider the expected field dependence of the isotope enrichment parameter α_1 (see Eq. (2.12)) for the triplet-born RPs

$$^T\alpha_1 = \frac{^Tp_g - {}^Tp_g^*}{1 - {}^Tp_g} , \qquad (2.76)$$

where ^Tp_g and $^Tp_g^*$ are the RP recombination probabilities before and after the substitution of magnetic isotopes for non-magnetic ones, respectively. Triplet-born RPs without magnetic nuclei cannot recombine so that $^Tp_g = 0$. As a result, $^T\alpha_1 = -{}^Tp_g^*$. This suggests that in the case under discussion the absolute value of the isotope enrichment parameter will pass through a maximum with the external magnetic field increasing.

The recombination probability of triplet-born RP cannot be simply identified as the probability of triplet-singlet conversion, $^Tp_g \neq p(T \to S)$. But the quantity $p(T \to S)$ highlights the special features of the RP's spin dynamics which play a decisive role in the recombination of RPs. It is precisely this quantity which determines the magnetic isotope effect in the recombination of RPs and indicates the range of variation in the RP recombination probabilities under the magnetic isotope substitution (see also [35]).

2.7 Kinetic equations for radical pair recombination

The probability of the RP recombination depends on many parameters: hfi, "in-cage" RP lifetime, radical reactivity, viscosity of a solvent, etc. Thus, in order to calculate the RP recombination probability one has to deal with spin dynamics, chemical transformations and molecular motion of the pair radicals. For this one should solve appropriate kinetic equations for the spin density matrix of the RPs. Comprehensive discussion of this subject is provided in [5], which will be reviewed very briefly here.

Several models exist, in which the diffusion of RP partners, their recombination and the escape process are treated in different manners. The choice of the appropriate model is determined, on the one hand, by the specific properties of the physical system under investigation and, on the other hand, by the convenience of mathematical treatment. The simplest phenomenological model is the so-called exponential one-position model, which takes into account the following processes: decomposition of the RPs into independent radicals, recombination of the RPs and S-T transitions. The kinetic equation for the spin density matrix ρ of the RPs is [5, 36]

$$\frac{\partial \rho}{\partial t} = \left(\frac{\partial \rho}{\partial t}\right)_{\text{spin dynamics}} - \frac{K_1(P\rho + \rho P)}{2} - \frac{\rho}{\tau} , \qquad (2.77)$$

where τ is the mean lifetime of the RPs, K_1 is the recombination rate constant of the RP in the reactive spin state, P is the projection operator into this state. The first term on the right hand side of Eq. (2.77) describes the spin evolution (see, e.g., Eq. (2.43)). Usually, recombination takes place in the singlet state. In this case P is the projection operator into the singlet state: $P = |S\rangle\langle S|$. The RP recombination probability equals

$$p_g = \int_0^\infty \frac{K_1(P\rho + \rho P)}{2} \, dt . \qquad (2.78)$$

The one-position model is often used for calculation of magnetic and spin effects in radical recombination reactions but it has an essential disadvantage [37]. Recombination as well as "escape" and S-T mixing all take place at the same time, at "one position". The physical dynamics of cage reactions is not adequately represented. In the course of random thermal motion in liquids the partners undergo consecutive approaches (re-encounters). For the time of contact the wavefunctions of the valence electrons overlap, the radicals are in the reaction regime and may recombine. During this time strong exchange interaction exists and the S and T states are far separated in energy. The lifetime of the RP in this reaction regime lies typically in the picosecond range. Hfi induced S-T transitions cannot develop. On the other hand, for the time intervals between re-encounters, the radicals are typically incapable of recombination but S-T transitions are effective. Although the one-position model includes the main processes in RPs and might be instructive

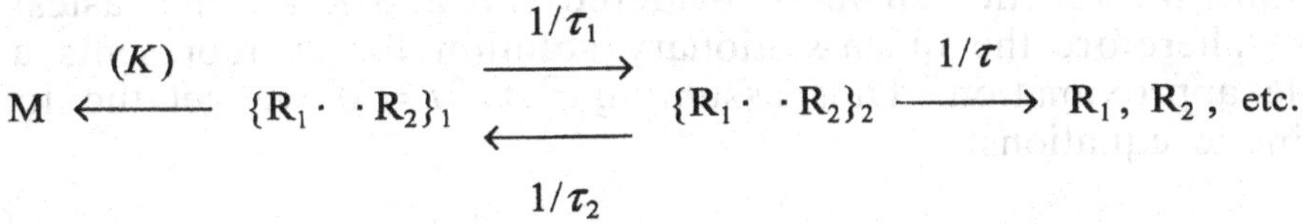

$$M \xleftarrow{\quad(K)\quad} \{R_1\cdot \ \cdot R_2\}_1 \underset{1/\tau_2}{\overset{1/\tau_1}{\rightleftarrows}} \{R_1\cdot \ \cdot R_2\}_2 \xrightarrow{\quad1/\tau\quad} R_1, R_2, \text{ etc.}$$

Fig. 2.10. Kinetic scheme of the two-position model. The rate constants are explained in the text.

for a qualitative discussion, it is not suited for a quantitative description of MIE.

More realistic is another phenomenological model, the exponential two-position model [34, 38]. Within this model RPs can be found in two positions: inside the reaction regime (position 1) and outside the reaction regime (position 2). In the first position RPs may either recombine or leave this position with a rate constant termed $1/\tau_1$. Within the non-reactive regime RPs are not able to recombine but S-T mixing occurs. With the rate constant $1/\tau_2$ RPs may return to position 1. Alternatively RPs may disappear from position 2: the radicals of the RP can separate to give independent radicals (diffusion of radicals into the bulk of a solvent), or they can convert to other, secondary, RPs, or they can react with acceptors. The total rate constant of these processes is $1/\tau$. This kinetic model is represented in Fig. 2.10.

For this reaction scheme the kinetic equations of the spin density matrix can be written down straightforwardly. If ρ_1 and ρ_2 are the spin density matrices of RPs in two positions, respectively, the kinetic equations are

$$\begin{cases} \dfrac{\partial \rho_1}{\partial t} = -\dfrac{K(P\rho_1 + \rho_1 P)}{2} - \dfrac{\rho_1}{\tau_1} + \dfrac{\rho_2}{\tau_2} + \left(\dfrac{\partial \rho_1}{\partial t}\right)_{\text{spin exchange}}, \\[2ex] \dfrac{\partial \rho_2}{\partial t} = \left(\dfrac{\partial \rho_2}{\partial t}\right)_{\text{spin dynamics}} + \dfrac{\rho_1}{\tau_1} - \dfrac{\rho_2}{\tau_2} - \dfrac{\rho_2}{\tau}. \end{cases} \tag{2.79}$$

The spin dynamics is described by an equation like Eq. (2.43). In the first position, strong Heisenberg exchange interaction acts. However it is well justified (see, e.g., [5]) to approximate the exchange dephasing effect with $(\rho_1)_{\text{ST}} = 0$. With this approximation the first equation in Eqs. (2.79) is replaced by the equations:

$$\begin{cases} \dfrac{\partial \rho_1}{\partial t} = -K P\rho_1 P - \dfrac{\rho_1}{\tau_1} + \dfrac{\rho_2}{\tau_2}, \\[2ex] (\rho_1)_{\text{ST}} = 0. \end{cases} \tag{2.80}$$

Within the kinetic scheme considered $1/\tau_1$ and K are the fastest processes. Therefore the quasi-stationary solution for ρ_1 represents a reasonable approximation. Thus assuming $\partial \rho_1/\partial t = 0$ we get the following kinetic equations:

$$
\begin{cases}
- K P \rho_1 P - \dfrac{\rho_1}{\tau_1} + \dfrac{\rho_2}{\tau_2} = 0 \ , \\[2mm]
(\rho_1)_{\text{ST}} = 0 \ , \\[2mm]
\dfrac{\partial \rho_2}{\partial t} = \left(\dfrac{\partial \rho_2}{\partial t} \right)_{\text{spin dynamics}} + \dfrac{\rho_1}{\tau_1} - \dfrac{\rho_2}{\tau_2} - \dfrac{\rho_2}{\tau} \ .
\end{cases}
\tag{2.81}
$$

The RP recombination probability is

$$
p_{\text{g}} = \int_0^\infty K P \rho_1 P \, dt \ .
\tag{2.82}
$$

The distribution of lifetimes in the position 2 between re-encounters is given by:

$$
f(t) = \frac{1}{\tau_2} \exp\left\{ -\left(\frac{1}{\tau_2} + \frac{1}{\tau} \right) t \right\} \ .
\tag{2.83}
$$

The total number of encounters equals

$$
n = \frac{\tau + \tau_2}{\tau_2} \ .
\tag{2.84}
$$

The mean time between re-encounters is

$$
\tau_{\text{m}} = \frac{\tau \, \tau_2}{\tau + \tau_2} \ .
\tag{2.85}
$$

And the total lifetime of RPs in the reaction regime is given by

$$
\tau_{\text{r}} = \tau_1 n \ .
\tag{2.86}
$$

Suppose RPs recombine from the singlet state. Then Eqs. (2.81) give:

$$\left\{ \begin{array}{l} \left(\dfrac{\partial \rho_2}{\partial t}\right)_{SS} = \left\{\left(\dfrac{\partial \rho_2}{\partial t}\right)_{\text{spin dynamics}}\right\}_{SS} - \left(\dfrac{1}{\tau_2}\dfrac{K\tau_1}{1+K\tau_1} + \dfrac{1}{\tau}\right)(\rho_2)_{SS} \ , \\[3mm] \left(\dfrac{\partial \rho_2}{\partial t}\right)_{ST} = \left\{\left(\dfrac{\partial \rho_2}{\partial t}\right)_{\text{spin dynamics}}\right\}_{ST} - \left(\dfrac{1}{\tau_2} + \dfrac{1}{\tau}\right)(\rho_2)_{ST} \ , \\[3mm] \left(\dfrac{\partial \rho_2}{\partial t}\right)_{TT'} = \left\{\left(\dfrac{\partial \rho_2}{\partial t}\right)_{\text{spin dynamics}}\right\}_{TT'} - \dfrac{1}{\tau}(\rho_2)_{TT'} \ . \end{array} \right. \qquad (2.87)$$

These equations contain only the density matrix of RPs in the position 2. In this approximation quantitative calculations of the RP recombination probability become simpler.

The most popular model for RPs, however, is the continuous diffusion model. Within this approach the RP state is represented by the partial density matrix $\rho(r, t)$ for a subensemble of pairs with a given distance r between the two partners. The time behaviour of $\rho(r, t)$ is governed by the RP spin dynamics and mutual diffusion of radicals. Suppose $\mathscr{H}(r)$ is the spin Hamiltonian of an RP with the partners separated by the distance r and D is the mutual diffusion coefficient, then the time dependence of $\rho(r, t)$ is determined by the equation

$$\frac{\partial \rho}{\partial t} = -\frac{i}{\hbar}[\mathscr{H}(r), \rho(r,t)] + D\,\Delta\rho(r,t) \ . \qquad (2.88)$$

The radicals are allowed to recombine below a distance b. Normally RPs recombine from their singlet state. Let K be the rate constant of singlet RP recombination at $b - a \le r \le b$, where a is a thickness of the reactive layer. Hence, at $r = b$ the following boundary conditions can be formulated

$$\left\{ \begin{array}{l} D\,\nabla\rho_{SS}(r,t)\big|_{r=b} = a\,K\,\rho_{SS}(b,t) \ , \\[2mm] D\,\nabla\rho_{TT'}(r,t)\big|_{r=b} = 0 \ . \end{array} \right. \qquad (2.89)$$

Exchange interaction dephases S and T states in the reaction regime so that the boundary condition for the non-diagonal elements of the density matrix can be approximated as

$$\rho_{ST}(r,t)\big|_{r=b} = 0 \ . \qquad (2.90)$$

The RP recombination probability is obtained as

$$p_g = 4\pi b^2 a K \int\limits_0^\infty \rho_{SS}(b,t)\,dt \ . \tag{2.91}$$

Using these kinetic equations the RP recombination probability, the time dependence of this recombination process, the variations with isotope substitution have been calculated and analyzed for some concrete systems (model and real). Some of these results will be presented in the next chapters.

3 Theoretical description of magnetic isotope effect in the Earth's magnetic field

Chemical reactions usually take place in the Earth's magnetic field which is about $5 \cdot 10^{-5}$ T. The contribution of the Zeeman interaction of unpaired electrons with the Earth's field to singlet-triplet evolution of short-lived RPs can be ignored with reasonable accuracy. Typically, the hfi in free radicals is an order of magnitude larger than the Zeeman interaction with the Earth's field. By virtue of this argument the results obtained for zero field can be applied to the theoretical description of MIE in the Earth's magnetic field. Thus, the Zeeman interaction of the RP's unpaired electrons will be ignored throughout this chapter.

This chapter outlines the results of the theoretical calculations of MIE in zero magnetic field for several model systems. We will discuss MIE dependence on RP parameters specifying their reactivity, molecular and spin motion. An extensive theoretical investigation of the recombination of neutral (uncharged) free radicals in zero magnetic field can be found in [38, 39].

3.1 Radical pairs with one magnetic nucleus (isotropic hyperfine coupling)

Let us start with considering the simplest model RP system where an hfi with only one magnetic nucleus may make an essential contribution to S-T conversion in RPs.

The RP recombination probability of a pair with one magnetic nucleus where isotropic hf coupling only contributes to S-T dynamics was expressed in the framework of the continuous diffusion model in [39]. In order to record results, we need to introduce several notations: K_S and K_T are the rate constants of the RP recombination in the reaction zone from the singlet and triplet states, respectively; $^Sp(S)$ and $^Tp(S)$ are the probabilities of the geminate RP recombination from the singlet state for the singlet and triplet precursors, respectively; the analogous probabilities for the recombination from the triplet state are $^Sp(T)$ and $^Tp(T)$. The total residence time of RPs in the reaction zone is τ_r, the overall number of encounters of an RP is n, and p_0 is the probability of the first encounter at the recombination radius. Using these notations the RP recombination probabilities are [39]

$$^Sp(S) = \frac{p_0 K_S \tau_r}{\Delta} \left[(1 + K_T \tau_r) \left(1 - \frac{w_0}{p_0} \right) + wn \right] \, ,$$

$$^Tp(S) = \frac{p_0 K_S \tau_r}{3\Delta} \left[(1 + K_T \tau_r) \frac{w_0}{p_0} + wn \right] \, ,$$

$$^Sp(T) = \frac{p_0 K_T \tau_r}{\Delta} \left[(1 + K_S \tau_r) \frac{w_0}{p_0} + wn \right] \, ,$$

$$^Tp(T) = \frac{2}{3} \frac{p_0 K_T \tau_r}{1 + K_T \tau_r} + \frac{p_0 K_T \tau_r}{3\Delta} \left[(1 + K_S \tau_r) \left(1 - \frac{w_0}{p_0} \right) + wn \right] \, , \qquad (3.1)$$

where

$$\Delta = (1 + K_S \tau_r)(1 + K_T \tau_r) + wn(2 + K_T \tau_r + K_S \tau_r) \, ,$$

$$w = \int_0^\infty p(S \to T; t) f(t) \, dt \, ,$$

$$w_0 = \int_0^\infty p(S \to T; t) f_0(t) \, dt \, .$$

These two integrals give the probability of the singlet $\to$ triplet transitions averaged over the RP lifetime distribution between re-encounters in the reaction zone (see, e.g., Eqs. (2.14) and (2.82)) w, while the same probability averaged over the RP lifetime distribution $f_0(t)$ before the first encounter of the RP partners at the reaction radius is w_0. For example, within the continuous diffusion model of RPs, $f_0(t)$ is given as (see [40])

$$f_0(t) = m_0 t^{-3/2} \exp \left(- \frac{\pi m_0^2}{p_0^2 t} \right) \, , \qquad (3.2)$$

where m_0 is expressed through the mutual diffusion coefficient $D = D_A + D_B$, the initial separation of radicals r_0 and the reaction radius b (see Eqs. (2.85) and also the discussion of this problem in [5])

$$m_0 = \frac{b(r_0 - b)}{2r_0 (\pi D)^{1/2}} \, . \qquad (3.3)$$

In the framework of the phenomenological two-position model, $f_0(t)$ is given by the same function $f(t)$ which describes the distribution of the re-encounters (see Eq. (2.82)). The probability of the first encounter of RPs at the reaction radius is equal to

$$p_0 = \frac{b}{r_0} \tag{3.4}$$

for the continuous diffusion model, and

$$p_0 = p = \frac{\tau}{\tau + \tau_2} \tag{3.5}$$

for the two-position model.

For the RPs with one magnetic nucleus of spin I the S $\rightarrow$ T dynamics originating from the isotropic hf interaction is given by the equation (note that Eqs. (2.48) and (2.50) give the results for $I = 1/2$ and $I = 1$ nuclear spins, respectively)

$$p(\text{S} \rightarrow \text{T}) = \frac{4I(I+1)}{1 + 4I(I+1)} \sin^2 \left(\frac{at\sqrt{1 + 4I(I+1)}}{4} \right) . \tag{3.6}$$

The average probabilities of S $\rightarrow$ T transitions are presented in Table 3.1 for two kinetic models of RPs, namely, the continuous diffusion and two-position models.

Equations (3.1) allow us to analyse MIE for the cases where only one magnetic nucleus plays a significant role in the S-T dynamics of RPs. Imagine an ensemble of RPs which possess no magnetic nuclei or only those with a negligible hfi, i.e., less than 0.1 mT. Now suppose that one of these nuclei is substituted by a magnetic isotope with spin I and strong hf coupling, and let us further assume that the hfi is the main mechanism of S-T conversion in RPs. This model situation is relevant to RPs formed, e.g., at the decomposition of fully deuterated DBK containing only ^{12}C isotope or containing ^{13}C isotope. Indeed, the hf cou-

Table 3.1. The average probabilities of singlet $\rightarrow$ triplet mixing for an RP with one magnetic nucleus.

Specification of quantity	Continuous diffusion model of an RP	Two-position model of an RP
wn	$k(|u|\tau_D/2)^{1/2}$	$k(\tau/\tau_2)\{u^2\tau_m^2/(1 + u^2\tau_m^2)\}$
w_0/p_0	$k\{1 - \cos(u\tau_0/2)^{1/2}\exp(-(|u|\tau_0/2)^{1/2})\}$	$ku^2\tau_m^2/(1 + u^2\tau_m^2)$

The parameters are: $k = 2I(I+1)/[1 + 4I(I+1)]$, $u^2 = a^2[1 + 4I(I+1)]/4$, $\tau_D = b^2/D$, $\tau_0 = (r_0 - b)^2/D$. In all these equations the isotropic hf coupling constant is quoted in the frequency units, rad/s.

pling with D in this case is about 0.1 mT and therefore it can be ignored. But when ^{12}C is substituted by ^{13}C in CO or CH_2 groups, strong hf coupling appears. Thus the results obtained for the model system with one magnetic nucleus can be used to analyse, e.g., the magnetic isotope effect at $^{12}C \rightarrow ^{13}C$ isotope substitution during the decomposition of fully deuterated DBK molecules. Here, the hfi is negligible before the isotope substitution, but after the isotope substitution a strong hfi with one nucleus is switched on.

Let us consider triplet-born geminate RPs which are permitted to recombine only from their singlet state. If all other mechanisms of S-T transitions (excluding hfi induced ones) are inefficient then before the isotope substitution triplet-born RPs will not recombine at all, so that $^Tp(S) = 0$. After the isotope substitution Eqs. (3.1) give us

$$^Tp(S) = \frac{p_0 K_S \tau_r}{3\Delta} \left(wn + \frac{w_0}{p_0} \right) ,$$

$$\Delta = 1 + K_S \tau_r + wn(2 + K_S \tau_r) . \tag{3.7}$$

For this model situation the MIE parameters are (see Eqs. (2.8) and (2.12))

$$^T\alpha = \frac{1}{1 - ^Tp(S)} , \tag{3.8}$$

$$^T\alpha_1 = -^Tp(S) . \tag{3.9}$$

According to these equations the MIE parameters depend on the following parameters: $a\tau_0$, which determines the S-T dynamics efficiency before the first contact at the reaction radius within a continuous diffusion model; $a\tau_D$ and $a\tau_m$, which determine the S-T dynamics efficiency between two subsequent contacts at the reaction radius within the continuous diffusion and two-position models, respectively; the parameter $K_S \tau_r$ (or $\lambda = K_S \tau_r/(1 + K_S \tau_r)$) which describes the chemical reactivity of the singlet RPs; and finally p_0 is the probability of the first encounter at the reaction radius.

For the model considered a maximum value of $^Tp(S)$ may reach 1/3. This limit can occur when radicals start from the reaction radius, i.e., $r_0 = b$, when the reactivity of the singlet RPs is high enough, i.e., $K_S \tau_r \gg 1$, and in addition when S-T mixing is particularly efficient, i.e., the condition $a\tau_D \gg 1$ is fulfilled. At these optimal conditions, the MIE parameters may approach the limiting values of

$$^T\alpha \rightarrow 1.5 , \tag{3.10}$$

$$^T\alpha_1 \rightarrow -0.333 . \tag{3.11}$$

One could expect that, with increasing $a\tau_D$ and $K_S\tau_r$, the RP recombination probability would tend towards one, $^Tp(S) \rightarrow 1$. In this case the isotope enrichment parameter $^T\alpha$ might be very large. However, this does not happen. As was discussed in the previous chapter, the total spin of electrons and nuclei in zero magnetic field has to be conserved when isotropic hf coupling alone is responsible for S-T conversion. As a result, not all triplet RPs can be converted to the singlet state. The numerical calculations show that the limits given by Eqs. (3.10) and (3.11) can only be reached with a very high viscosity of about 1000 cP or by radicals with high hf coupling constants, i.e., about 10–100 mT. In general, the probability for the RP recombination considered in the framework of the continuous diffusion model contains an oscillating term reflecting the S-T dynamics before the first contact at the reaction radius. However, these oscillations can only be observed with difficulty for the typical values of the RP parameters. Some results of numerical calculations are shown in Figs. 3.1–3.4.

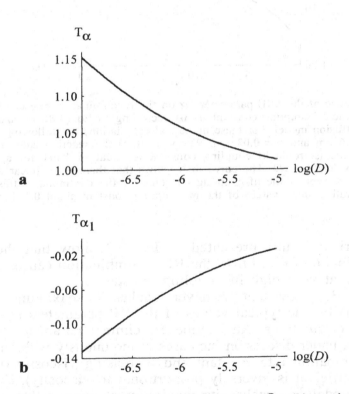

Fig. 3.1. Dependence of the MIE parameters $^T\alpha$ (a) and $^T\alpha_1$ (b) on the diffusion coefficient D according to Eqs. (3.8) and (3.9) calculated within the continuous diffusion model (D in cm^2/s units). For these numerical calculations the following parameters were used: the isotropic hf coupling constant in the RP which contains a magnetic isotope is equal to 1 mT, $b = 0.6$ nm, and $a = 0.03$ nm. RPs start from the reaction zone, i.e., $r_0 = b$, $K_S = 100$ ns^{-1}, and $I = 1/2$.

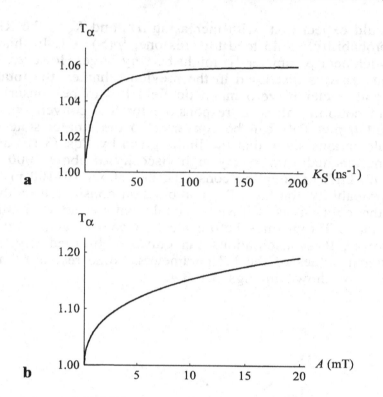

Fig. 3.2. Dependence of the MIE parameter $^T\alpha$ on the recombination rate constant K_S (**a**) and on the isotropic hf coupling constant A (**b**) according to Eq. (3.8), calculated within the continuous diffusion model. For these numerical calculations the following parameters were used: $b = 0.6$ nm and $a = 0.03$ nm. RPs start from the reaction zone, i.e., $r_0 = b$. $D = 10^{-6}$ cm²/s, the isotropic hf coupling constant is equal to 1 mT for **a**, and $K_S = 1000$ ns^{-1} for **b**, and $I = 1/2$. The curve in **b** shows that there is no linear dependence of $^T\alpha$ on the square root of the hf coupling constant for the continuous diffusion model even for rather small values of the hf coupling constant about 0.1–1 mT.

The numerical results presented in Fig. 3.2 show that the largest possible contribution of an hfi to the RP recombination can be expected to appear only at very high hf coupling constants.

Figures 3.1–3.4 show that $^T\alpha$ never reaches its maximum value 1.5 (see Eq. (3.10)) for the typical values of the RP parameters used during the numerical calculations. An isotope enrichment parameter $^T\alpha$ in the model situation under discussion increases monotonously with an increasing RP recombination rate constant and increasing viscosity (note that a diffusion coefficient is inversely proportional to viscosity). Within the two-position model $^T\alpha$ reaches its maximum at some value of a lifetime τ_2. But as one can see from Fig. 3.4a and b this extremum is not very well pronounced. Generally, an extremum is expected in a viscosity dependence of MIE. One reason why this behaviour did not manifest itself in Fig. 3.1 and was not well pronounced in Fig. 3.4 may be

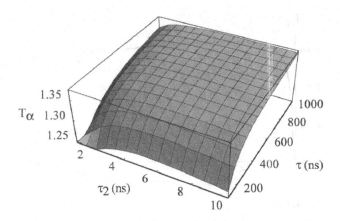

Fig. 3.3. The expected values of the MIE parameter $^T\alpha$ (Eq. (3.8)) calculated within the two-position model of RPs: a general view of the $^T\alpha$ behaviour as the function of the lifetime τ_2 between two subsequent re-encounters of partners in RPs and the lifetime τ which determines the escape from the RP state. Other parameters are: $K_S = 1000$ ns^{-1}, $A = 1$ mT, $\tau_r = 0.01$ ns, and $I = 1/2$.

due to the fact that, in the model situation under discussion, any S-T transitions were fully ignored before an isotope substitution. At high viscosity (long lifetime between subsequent encounters of RPs) even relatively small hfi are able to mix S and T states substantially. As a result, at very high viscosity both RPs, before and after isotope substitution, can recombine with practically equal probabilities, and MIE is expected to vanish. When viscosity tends towards zero, the geminate RP recombination probability becomes zero also. Thus, as a function of viscosity η (or the diffusion coefficient D, $D \propto 1/\eta$), $^T\alpha$ should pass through an extremum value. To illustrate this feature of MIE let us consider MIE at H $\rightarrow$ D isotope substitution for an RP which possesses only one magnetic nucleus. RP recombination probabilities for both pairs can be found using Eqs. (3.7). In the case of protonated RPs the nucleus of spin $I = 1/2$ has the isotropic hf coupling constant a_H while deuterated pairs possess nuclear spin $I = 1$ with $a_D = 0.1535\,a_H$. In this case, the MIE parameters are chosen as (see Eqs. (2.8) and (2.12))

$$^T\alpha = \frac{1 - {}^Tp_H(S)}{1 - {}^Tp_D(S)} \, , \tag{3.12}$$

$$^T\alpha_1 = \frac{{}^Tp_H(S) - {}^Tp_D(S)}{1 - {}^Tp_H(S)} \, . \tag{3.13}$$

Substituting the recombination probabilities given in Eq. (3.13) for those given by Eq. (3.7) we obtain

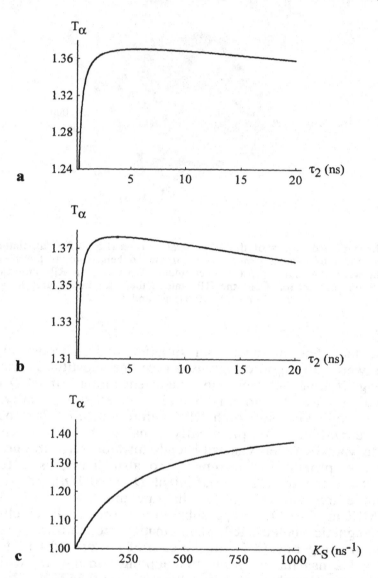

Fig. 3.4. The expected values of the MIE parameter $^T\alpha$ (Eq. (3.8)) within the two-position model of RPs. **a** and **b** Dependence of $^T\alpha$ on τ_2 for $I = 1/2$ and $I = 1$, respectively. **c** Dependence of $^T\alpha$ on a RP recombination rate constant. The parameters used during these numerical calculations are: $\tau = 1000$ ns, $\tau_r = 0.01$ ns, $A = 1$ mT, $K_S = 1000$ ns^{-1} for **a** and **b**, and $\tau_2 = 5$ ns, $I = 1$ for **c**.

$$
^T\alpha_1 = \frac{w_H n - w_D n}{3(3 + 3K_S\tau_r + 6w_H n + 2w_H n K_S\tau_r)}
$$

$$
\times \frac{K_S\tau_r(1 + K_S\tau_r)}{1 + K_S\tau_r + 2w_D n + w_D n K_S\tau_r} , \tag{3.14}
$$

where $w_H n$ and $w_D n$ characterize the efficiency wn of S-T mixing in the protonated and deuterated RP, respectively (see Table 3.1). For example, in the case of the continuous diffusion model, $w_H n$ and $w_D n$ are rendered as

$$w_H n = \frac{3}{8\sqrt{2}} \left(|a_H| \tau_D \right)^{1/2} ,$$

$$w_D n = \frac{2}{3\sqrt{3}} \left(|a_D| \tau_D \right)^{1/2} . \tag{3.15}$$

The quantities $w_H n$ and $w_D n$ are proportional to $\eta^{1/2}$ or $D^{-1/2}$. Taking this into account we obtain the following estimates:

$$^T\alpha_1 \approx (w_H n - w_D n) \frac{K_S \tau_r}{9} \propto \eta^{3/2} \propto D^{-3/2} \tag{3.16}$$

in low viscosity case when $w_H n$, $w_D n < 1$, $K_S \tau_r < 1$, and

$$^T\alpha_1 \approx \frac{w_H n - w_D n}{6 w_H n\, w_D n} \propto \eta^{-1/2} \propto D^{1/2} \tag{3.17}$$

if the viscosity is high enough to provide the conditions $w_H n$, $w_D n > 1$, and $K_S \tau_r > 1$. Thus, with increasing viscosity (decreasing diffusion coefficient) $^T\alpha_1$ should pass through a maximum. To detect this maximum, the viscosity must be high enough so that S-T mixing can be efficient in *both* RPs, before and after an isotope substitution. The largest MIE is expected for a viscosity when S-T mixing is not pronounced during an RP in-cage lifetime in deuterated RPs while the hf coupling in protonated pairs produces an efficient S-T conversion. Therefore, the maximum of $^T\alpha_1$ should appear under the following conditions

$$|a_H| \tau_D > 1 , \qquad |a_D| \tau_D < 1 . \tag{3.18}$$

For the typical values of the proton isotropic hf coupling constants about 1 mT, Eq. (3.18) predicts the extremum for $^T\alpha_1$ at high viscosities only, i.e., when the diffusion coefficient becomes $D \approx 10^{-7}$–10^{-8} cm^2/s.

In many cases reactivity of free radicals is high enough to correspond to a limit $K_S \tau_r > 1$. Under this condition, we find that

$$^T\alpha_1 \approx \frac{w_H n - w_D n}{3(3 + 2 w_H n)(1 + w_D n)} . \tag{3.19}$$

According to this expression, $^T\alpha_1$ first increases in proportion to $\eta^{1/2}$ and then decreases as $\eta^{-1/2}$ with increasing viscosity. The largest value for $^T\alpha_1$ is expected at

$$\tau_D = \frac{6^{3/2}}{\gamma_e \sqrt{A_H A_D}} \approx \frac{36}{\gamma_e A_H} \, , \tag{3.20}$$

where A_H and A_D are the hf coupling constants quoted in the magnetic field units. For the coupling constants about 1 mT, τ_D has to be about 200 ns, which requires a rather high viscosity of the solvent. These qualitative conclusions are illustrated by the numerical calculations (see Figs. 3.5 and 3.6).

We see from Fig. 3.5 that MIE parameters reach the extrema but only at very high viscosity when $D \approx 3 \cdot 10^{-8}$ cm^2/s. Figure 3.6 shows that there is an optimum value of an hf coupling constant for the MIE. However, the remarkable feature of the data in Figs. 3.5 and 3.6 is that a very small MIE, i.e., only a few percent, is predicted. (Note that without any S-T mixing according to Eqs. (3.12) and (3.13) $^T\alpha = 1$ and $^T\alpha_1 = 0$.) This surprisingly small MIE reflects the peculiar situation of the distribution of a lifetime between two subsequent in-cage RP encounters within the continuous diffusion model (see Eq. (2.14) and Fig. 2.5): the distribution decreases very slowly with increasing lifetime, and

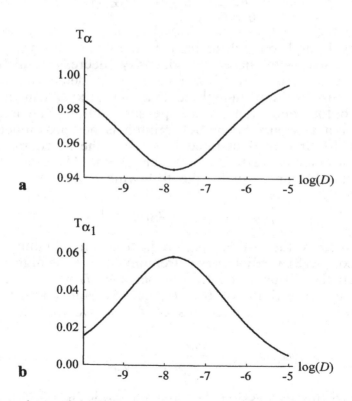

Fig. 3.5. Dependence of the MIE parameters $^T\alpha$ (**a**) and $^T\alpha_1$ (**b**) on the diffusion coefficient D (D in cm^2/s units). The parameters used for these calculations are: $b = 0.6$ nm, $a = 0.03$ nm, $K_S = 100$ ns^{-1}, $r_0 = b$, and $A_H = 1$ mT.

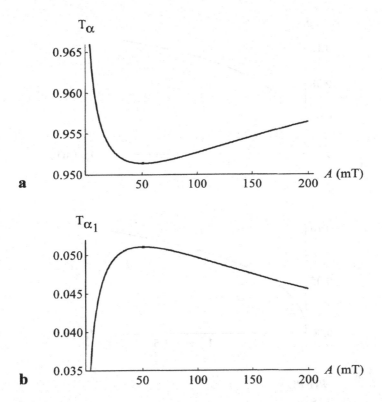

Fig. 3.6. Dependence of the MIE parameters $^T\alpha$ (**a**) and $^T\alpha_1$ (**b**) on the hf coupling constant A in the protonated RPs during the H → D isotope substitution. In the deuterated RPs the hf coupling constant is reduced 6.51 times. The parameters used for these calculations are: $D = 10^{-6}$ cm²/s, $b = 0.6$ nm, $a = 0.03$ nm, $r_0 = b$, and $K_S = 100$ ns⁻¹ (the same as for Fig. 3.5).

a large part of the re-encounters in a homogeneous solution occurs after a long diffusion time. That means that even for RPs with a slow (but not exactly zero!) S-T mixing there will be some re-encounters, after a sufficiently long diffusion time, demonstrating an efficient S-T conversion.

In the model RP considered, for both protonated and deuterated pairs, the hfi induces S-T transitions, so that a long diffusion time will favour efficient singlet-triplet mixing for RPs with both isotope decompositions. As a result, RP recombination probabilities for the protonated and deuterated systems are expected to differ less than the ratio of their nuclear magnetic moments. Figure 3.7 illustrates this feature. The results exposed confirm our hypothesis that after an isotope substitution RP recombination probabilities change less than a corresponding change of the hyperfine coupling constant.

The contribution to RP recombination from those pairs which re-encounter each other after a long time can be reduced due to the fact

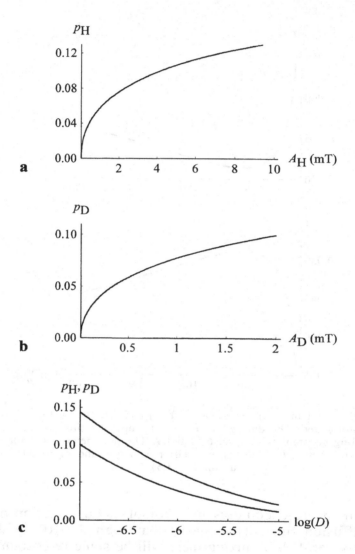

Fig. 3.7. Dependence of RP recombination probabilities for protonated (p_H) and deuterated (p_D) triplet-born pairs on the corresponding hf coupling constant for proton (A_H) and deuteron (A_D) and on the diffusion coefficient D. Parameters of RPs are: $b = 0.6$ nm, $a = 0.03$ nm, $r_0 = b$, $K_S = 1000$ ns^{-1}, and $A_H = 2.272$ mT in **c**; and $D = 10^{-6}$ cm^2/s in **a** and **b**. In **c** (p_H, top curve; p_D, bottom curve) D changes in the interval 10^{-7}–10^{-5} cm^2/s.

that radicals react with acceptors or undergo some other chemical reactions. In the presence of radical acceptors RP recombination probability is described by Eqs. (3.1) as well. The only difference is that parameters τ_r and wn are modified. For example, in the framework of the continuous diffusion model, parameters τ_r and wn are equal to [39] (compare with Table 3.1):

$$\tau_r = \frac{abD^{-1}}{1 + \sqrt{K_a \tau_D}} ,$$

$$wn = k \frac{\sqrt{\left(K_a + \sqrt{K_a^2 + u^2}\right)\dfrac{\tau_D}{2}} - \sqrt{K_a \tau_D}}{1 + \sqrt{K_a \tau_D}} , \qquad (3.21)$$

where K_a is a rate for the reaction with acceptors. However, acceptors reduce RP recombination probabilities for both isotope compositions. Thus, an increase of the magnetic isotope effect by means of an addition of radical acceptors is not a promising way.

To increase MIE, it seems to be more favourable to proceed with chemical reactions in non-homogeneous solutions. For instance, in micelle the statistics of re-encounters can be changed drastically [17]. For RP recombination in micelle, it might be reasonable to use the two-position phenomenological model. Within this model larger MIE can be obtained. Figure 3.8 shows the results of the numerical calculations of the MIE parameters defined by Eqs. (3.12) and (3.13) for the two-position model. The most important feature of results presented in Fig. 3.8 is that in the framework of the two-position model MIE parameters vary about 0.15–0.2, i.e., more than in the case of the continuous diffusion model. Another remarkable feature of the two-position model is that in order to observe MIE there has to be an optimal lifetime between subsequent re-encounters of two RP partners and an optimal hf coupling (see Fig. 3.8).

Similar arguments concerning MIE can be put forward for the recombination of singlet-born RPs. For example, for RPs starting from the reaction radius, $r_0 = b$, Eqs. (3.1) lead to

$$^S p^*(S) = \frac{K_S \tau_r (1 + wn)}{\Delta} ,$$

$$\Delta = 1 + K_S \tau_r + wn(2 + K_S \tau_r) . \qquad (3.22)$$

Without a magnetic nucleus, we have

$$^S p(S) = \frac{K_S \tau_r}{1 + K_S \tau_r} . \qquad (3.23)$$

The isotope enrichment parameters α_1 (see Eqs. (2.23) and (2.24)) for singlet-born and triplet-born RPs are connected by the relation

$$^S \alpha_1 = -3 \, ^T \alpha_1 , \qquad (3.24)$$

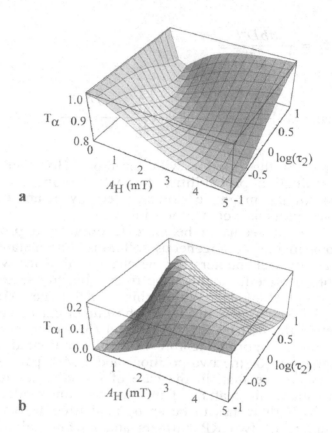

Fig. 3.8. A visualization of the MIE parameters $^{T}\alpha$ (**a**) and $^{T}\alpha_1$ (**b**) during H → D isotope substitution as depending on the hf coupling constant in protonated RPs and the lifetime between two subsequent re-encounters at the reaction regime τ_2 (τ_2 in ns units). The parameters used for these calculations are: $\tau = 100$ ns, $\tau_r = 0.01$ ns, and $K_S = 1000$ ns^{-1}.

where according to Eq. (3.9) $^{T}\alpha_1 = -^{T}p(S)$. Equation (3.24) shows clearly that at least for this model situation, the triplet precursors have no advantages compared to the singlet precursors from the point of view of optimal conditions for MIE.

When both RPs, before and after an isotope substitution, possess one magnetic nucleus, for example, proton H and deuteron D, then an isotope enrichment parameter for triplet-born RPs is given by Eq. (3.14) while for the singlet-born RPs it is

$$^{S}\alpha_1 = -\frac{(w_{H}n - w_{D}n)K_{S}\tau_r}{(1 + 2w_{H}n)(1 + K_{S}\tau_r + 2w_{D}n + w_{D}n K_{S}\tau_r)} . \qquad (3.25)$$

If $K_{S}\tau_r \gg 1$, this expression is reduced to

$$^{S}\alpha_1 \approx -\frac{w_H n - w_D n}{(1 + 2w_H n)(1 + w_D n)} . \tag{3.26}$$

As is expected, $^{S}\alpha_1$ and $^{T}\alpha_1$ have opposite signs. With increasing viscosity $^{S}\alpha_1$ passes through a minimum, and $^{S}\alpha_1$ depends on $K_S \tau_r$ monotonously.

3.2 Radical pairs with magnetically equivalent nuclei (isotropic hyperfine coupling)

Free radicals may possess several nuclei which are magnetically equivalent. For example, there are nine equivalent protons in the free radical $C(CH_3)_3$. This radical is formed, e.g., during the photolysis of di-tert-butyl ketone [41]

$$(CH_3)_3CCOC(CH_3)_3 \longrightarrow (CH_3)_3CCO + C(CH_3)_3 . \tag{3.27}$$

For this particular example, the contribution of the proton isotropic hf coupling in $(CH_3)_3CCO$ to S-T dynamics is negligible while the hf coupling in the partner $C(CH_3)_3$ may produce an efficient S-T conversion. Keeping in mind this type of RP let us consider a model RP with one radical having any number of magnetically equivalent nuclei, while its partner exhibits a negligible hf coupling. The recombination probability of this model RP can be calculated in the following way [39]. The pairs can be divided into subensembles, each having only one magnetic nucleus with the spin equal to one of the possible values of the total spin of the equivalent nuclei. For each subensemble of RPs, the recombination probability can be calculated by using the results described in the previous section. For instance, let one radical of a pair have three equivalent protons. The total nuclear spin equals 1/2 or 3/2. Taking into account the statistical weights of RPs with equal total nuclear spins, the recombination probability for RP with three equivalent protons is expressed as

$$p = \frac{1}{2}\left[p\left(\frac{1}{2}; a_H\right) + p\left(\frac{3}{2}; a_H\right) \right], \tag{3.28}$$

where $p(I, a)$ is the recombination probability for a one-nuclear RP having spin I and an hf coupling constant a. If all three protons are substituted by deuterons, the total spin of D nuclei becomes 3, 2, 1, or 0 with statistical weights of 7/27, 10/27, 9/27, and 1/27, respectively. The recombination probability of an RP with a deuterated radical is calculated from the following formula

$$p = \frac{1}{27}\left[7p(3; a_D) + 10p(2; a_D) + 9p(1; a_D) + p(0; a_D)\right] . \quad (3.29)$$

The difference in p calculated from Eqs. (3.27) and (3.28) determines the magnetic isotope effect. The RP recombination probabilities and the MIE parameters for other RPs with magnetically equivalent nuclei can be calculated in a similar manner using the summation rule of nuclear spin moments and the results of the above section. This means that at isotope substitution of all magnetically equivalent nuclei, the MIE parameters will show a behaviour very similar to that of one-nuclear pairs.

3.3 Radical pairs with many magnetically non-equivalent nuclei (isotropic hyperfine coupling)

The singlet-triplet dynamics in RPs with many magnetically non-equivalent nuclei can be considered in the framework of the semiclassical description of the hyperfine interaction. Here, classical vectors are substituted for nuclear spin moment operators in the spin Hamiltonian of hfi [33]. Using this approach, the RP recombination probability in zero field was determined (see [39]). The recombination probabilities $^Sp(S)$, $^Tp(S)$, $^Sp(T)$ and $^Tp(T)$ were obtained by a summation of the contributions of all re-encounters of radicals at the reaction radius. Here, we present only the final results for several cases. Let us take RPs where the radical motion fits the continuous diffusion model. Assume that there are no acceptors of radicals, and the radicals of a pair are in contact at the generation moment. Under these circumstances, the RP recombination probabilities are given by the equations

$$^Sp(S) = K_S\tau_r \frac{1 + K_T\tau_r + f}{\Delta} ,$$

$$^Tp(S) = \frac{K_S\tau_r f}{\Delta} ,$$

$$^Sp(T) = \frac{3K_T\tau_r f}{\Delta} ,$$

$$^Tp(T) = K_T\tau_r \frac{1 + K_S\tau_r + 3f}{\Delta} , \quad (3.30)$$

where

$$\Delta = (1 + K_S\tau_r)(1 + K_T\tau_r) + f(4 + 3K_T\tau_r + K_S\tau_r) ,$$

$$f = 0.145\sqrt{a_{ef}\tau_D} ,$$

$$\sqrt{a_{ef}} = \frac{1}{3}\left[\sqrt{a_{1,ef}} + \sqrt{a_{2,ef}} + 2(a_{1,ef}^2 + a_{2,ef}^2)^{1/4}\right] .$$

In these expressions all parameters coincide with those in Eqs. (3.1), while a_{1ef} and a_{2ef} are the effective hf coupling constants of two RP partners quoted in the frequency units rad/s. The H $\rightarrow$ D isotope substitution, for example, reduces the effective hfi parameter a_{ef} four times. In contrast to the previously considered models of RPs with one magnetic nucleus or with many magnetically equivalent nuclei, the hfi in the present RP model is able to convert all triplet-born RPs into the singlet state, and as a result, triplet-born RPs may, in general, recombine with the probability 1. This result confirms our statements in Sect. 2.6. It stems from the fact that for RPs with many magnetically non-equivalent nuclei the total spin conservation rule at zero magnetic field creates much fewer restrictions for T $\rightarrow$ S transitions than in the case of RPs with one or many magnetically equivalent nuclei.

To illustrate the potential possibilities of the model under discussion let us consider the variation of the RP recombination probability and the MIE parameters at the H $\rightarrow$ D isotope substitution in the case of triplet-born pairs which are allowed to recombine only from their singlet state. Under these circumstances RP recombination probability increases monotonously with the reactivity to recombine and the effective hf coupling constant. The results of the numerical calculations using Eqs. (3.30), and Eqs. (3.12) and (3.13) are depicted in Figs. 3.9 and 3.10.

From Fig. 3.9 it follows that the MIE parameters $^T\alpha$ and $^T\alpha_1$ and the RP recombination probabilities vary noticeably when the proton effective hf coupling constant changes in the interval 0.1–100 mT.

Figure 3.10 shows that in the situation considered the influence of the diffusion coefficient on the MIE parameters α and α_1 and the RP recombination probabilities is more pronounced than for the one-nucleus RPs and RPs with equivalent nuclei (compare, for instance, Figs. 3.5 and 3.10).

According to Eqs. (3.30), for RPs with many magnetically non-equivalent nuclei MIE parameters are monotonously dependent on an effective hf coupling constant and a diffusion coefficient. For triplet-born RPs with many magnetically non-equivalent nuclei, the MIE parameters increase steadily with an effective hf coupling constant and a diffusion coefficient. At the same time the MIE parameters for singlet-born RPs defined by the equations analogous to Eqs. (3.12) and (3.13) may demonstrate a non-monotonous behaviour with respect to variation of an effective hf coupling constant and a molecular mobility. Figures 3.10–3.12 illustrate the dependence of the MIE parameters on an RP precursor multiplicity. These numerical calculations were done using the RP recombination probabilities given by Eqs. (3.30) assuming that RPs recombine only from their singlet state.

Two observations that follow from Figs. 3.10–3.12 can be made. In the case of the recombination of singlet-born RPs, there are optimal values of a diffusion coefficient and of an hf coupling constant with respect to MIE parameters. However, these optimal conditions require a

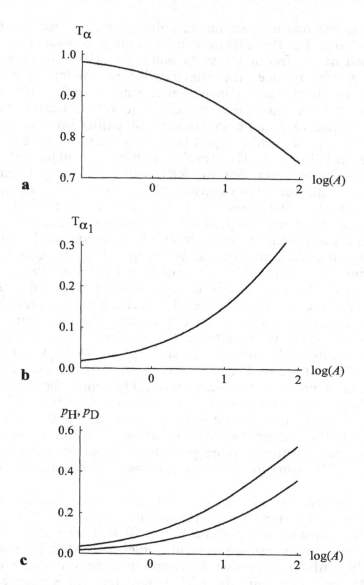

Fig. 3.9. Dependence of the MIE parameters $^T\alpha$ (**a**) and $^T\alpha_1$ (**b**), and of the recombination probabilities for protonated (p_H, top curve) and deuterated (p_D, bottom curve) triplet-born RPs (**c**) on the effective hyperfine coupling constant A (A in mT units). The parameters used for these calculations are: $K_S = 1000$ ns^{-1}, $D = 10^{-6}$ cm^2/s, $b = 0.6$ nm, and $a = 0.03$ nm. For all these curves A is given for fully protonated RPs. This presentation allows one to compare directly results for two counterpart RPs during the H $\rightarrow$ D isotope substitution.

rather high solvent viscosity. For instance, the extremum in Fig. 3.11 appears only at $D \approx 3 \cdot 10^{-8}$ cm^2/s. Another interesting point is that for both singlet-born and triplet-born RPs the scales of MIE parameters are comparable.

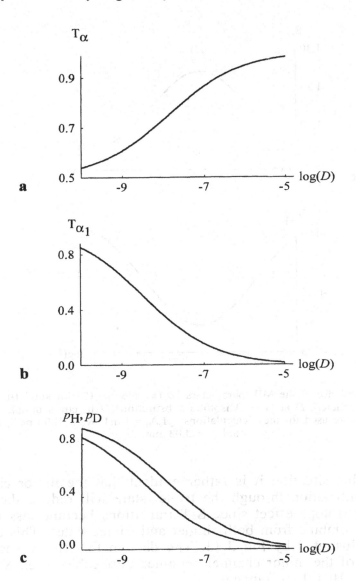

Fig. 3.10. Dependence of the MIE parameters $^T\alpha$ (**a**) and $^T\alpha_1$ (**b**), and of the recombination probabilities for protonated (p_H, top curve) and deuterated (p_D, bottom curve) RPs (**c**) on the diffusion coefficient D (D in cm^2/s units). The parameters used for these calculations are: $K_S = 1000$ ns^{-1}, $(A_{ef})_H = 1$ mT, $b = 0.6$ nm, and $a = 0.03$ nm.

A subject of great importance is the possibility of RPs to recombine not only from their singlet state but also from the triplet state. Suppose that the recombination from the singlet state is the major channel of recombination and that the triplet pairs are able to recombine as well but with much less efficiency (the minor channel of recombination). Let us further assume that both reaction channels give the same prod-

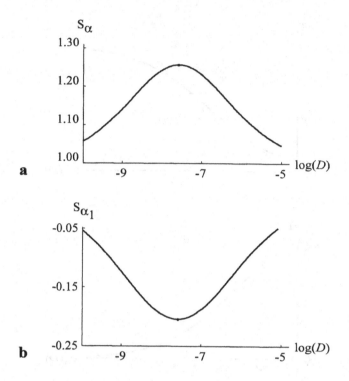

Fig. 3.11. Dependence of the MIE parameters $^S\alpha$ (**a**) and $^S\alpha_1$ (**b**) for singlet-born pairs on the diffusion coefficient D at H $\rightarrow$ D isotope substitution (D in cm^2/s units). The following parameters were used for these calculations: $(A_{ef})_H = 1$ mT, $K_S = 1000$ ns^{-1}, $b = 0.6$ nm, and $a = 0.03$ nm.

uct. Under this situation it is rather evident that the minor channel of the RP recombination through the triplet state will reduce the scale of the magnetic isotope effect since S-T transitions become less important when RPs recombine from both singlet and triplet states. This behaviour of MIE is illustrated by Fig. 3.13. The data in this figure show that a contribution of the minor channel becomes noticeable when $K_T\tau_r \gtrsim 0.1$. When $K_T\tau_r > 10$ MIE disappears.

The RP recombination through the minor channel also affects the diffusion coefficient dependence of the MIE parameters $^T\alpha$ and $^T\alpha_1$ as defined by Eqs. (3.12) and (3.13) for triplet-born RPs at H $\rightarrow$ D isotope substitution: the functions $^T\alpha(D)$ and $^T\alpha_1(D)$ may exhibit the extreme, but only at a rather large viscosity (see Fig. 3.14, compare with Fig. 3.10). The decrease of the MIE at a large viscosity (or at a small diffusion coefficient) is explained in the following way. At a large viscosity, an RP lifetime in a reaction zone is long enough so that not only the singlet, but also the triplet RPs succeed in recombining during their collision at the reaction radius. Therefore, at a large viscosity of solvents the singlet-triplet transitions become unimportant for

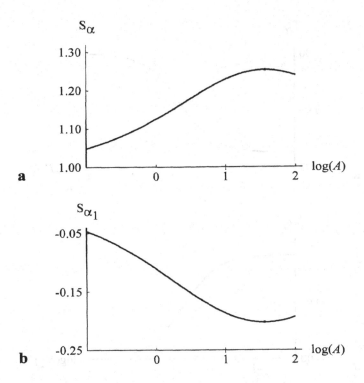

Fig. 3.12. Dependence of the MIE parameters $^S\alpha$ (**a**) and $^S\alpha_1$ (**b**) for singlet-born pairs on the proton hyperfine coupling constant A at H → D isotope substitution (A in mT units). The following parameters were used for these calculations: $D = 10^{-6}$ cm^2/s, $K_S = 1000$ ns^{-1}, $b = 0.6$ nm, $a = 0.03$ nm. These curves should be compared with Fig. 3.9.

the RP recombination if recombination from singlet and triplet states gives the same final product. Thus, the MIE parameters have to diminish when, for the RP recombination via the minor channel, this condition

$$K_T \tau_r \geq 1 \tag{3.31}$$

is fulfilled. Figure 3.14 illustrates that there is an optimal with respect to the MIE value of a diffusion coefficient.

The approach outlined in this section is well justified when magnetically non-equivalent nuclei possess comparable hf coupling constants. Experimental systems may not match this requirement. The magnetic ^{13}C enrichment observed at dibenzyl ketone (DBK) photolysis [20, 42] can serve here as a well-studied illustration. The DBK decomposition occurs from the triplet state and gives, e.g., RPs with ^{13}C or ^{12}C in the CO group:

$$PhCH_2{}^{12}CO\cdot \ \cdot PhCH_2 , \qquad RP\ I ,$$

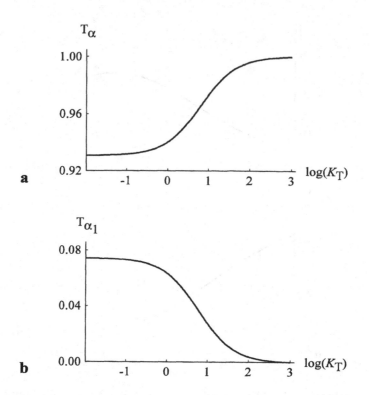

Fig. 3.13. The influence of the minor recombination channel through the triplet state on the MIE parameters $^T\alpha$ (**a**) and $^T\alpha_1$ (**b**) as defined by Eqs. (3.12) and (3.13) for triplet-born RPs at H → D isotope substitution. Note that K_T is the rate of the recombination of RPs from the triplet state. In this diagram K_T changes in the interval 10^7-10^{12} s^{-1}. The parameters used for these calculations are: $(A_{ef})_H = 2$ mT, $D = 10^{-6}$ cm^2/s, $K_S = 1000$ ns^{-1}, $b = 0.6$ nm, and $a = 0.03$ nm. It is supposed here that the minor channel of the RP recombination gives the same final product as the major channel.

$$PhCH_2{}^{13}CO\cdot\;\cdot PhCH_2\;, \qquad RP\;II\;. \qquad\qquad (3.32)$$

An appreciable ^{13}C enrichment in the initial ketone has been experimentally measured. The effect is associated with a strong hfi with ^{13}C in RP II namely $A_C = 12.5$ mT [30]. The hf coupling constants with protons in benzyl radical are of an order of magnitude smaller than A_C. To interpret theoretically the experimental data on MIE, one should calculate the recombination probabilities of RP I and RP II Eq. (3.32). Strictly speaking, the semiclassical model described above is inapplicable to calculations of the RP II recombination probability. The semiclassical theory of recombination of RPs with many magnetically non-equivalent nuclei employs the Gaussian distribution of local hfi fields in radicals. For the radical PhCH$_2{}^{13}$CO this supposition does not hold, since here the hfi with H and with ^{13}C differs by several orders of magnitude

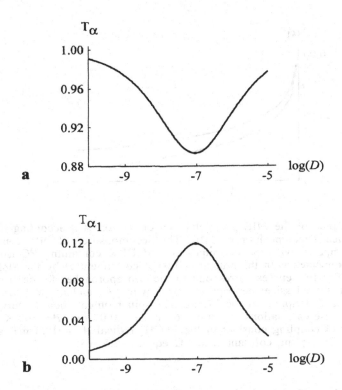

Fig. 3.14. Dependence of the MIE parameters $^{\mathrm{T}}\alpha$ (a) and $^{\mathrm{T}}\alpha_1$ (b) for triplet-born RPs on the diffusion coefficient D at the H $\rightarrow$ D isotope substitution (D in cm^2/s units). The parameters used for these calculations are: $K_{\mathrm{S}} = 1000$ ns^{-1}, $K_{\mathrm{T}} = 0.5$ ns^{-1}, $(A_{\mathrm{ef}})_{\mathrm{H}} = 2$ mT, $b = 0.6$ nm, and $a = 0.03$ nm.

($A_{\mathrm{H}} < 0.04$ mT [30]). As a result of this ratio of the hf coupling constants, the total local magnetic field, induced by magnetic nuclei, disobeys the applicability requirement of the central limit theorem of the probability theory and thus, the local field distribution cannot be treated as Gaussian (see [33]) in the particular radical considered. Hence it was necessary to verify the possibility of applying the semiclassical description of the hfi for an analysis of the MIE parameters when the RP has a magnetic nucleus with an hf coupling constant greatly differing from those of other nuclei in the pair.

With this aim the recombination of a model pair, with one radical having many magnetically non-equivalent nuclei and the other radical possessing only one nucleus with an $I = 1/2$ spin was investigated in [39]. The hfi induced spin dynamics in the latter radical was described in the proper quantum-mechanical way, and that in the radical which possesses many magnetically non-equivalent nuclei was considered in the semiclassical approximation. The problem was reduced to the numerical solution of eight algebraic linear equations with constant coefficients.

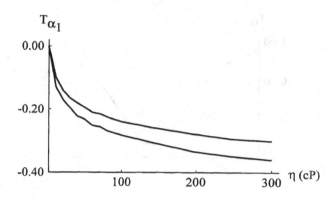

Fig. 3.15. Dependence of the MIE parameter $^T\alpha_1$ on viscosity η according to Eq. (3.33) for the photochemical decomposition of DBK. The decomposition of DBK containing only ^{12}C isotopes is compared with the decomposition of DBK containing ^{13}C isotope in the CO group. η is connected with the mutual diffusion coefficient D by the Stokes relation $D = 2kT/3\pi b\eta$. The two curves correspond to the two approaches for describing the RP spin dynamics: the semiclassical description of hfi with ^{13}C nucleus (top curve), and the quantum mechanical description of the ^{13}C nuclear spin moment (bottom curve). The parameters used for these calculations are: $b = 0.6$ nm, $a = 0.03$ nm, $T = 300$ K, $K_S = 1000$ ns^{-1}, the effective hf coupling constant in the PhCH$_2$ radical $A_{ef} = 1.23$ mT, and hf coupling constant with ^{13}C equals 12.5 mT.

Using the results of calculations in [39] the MIE parameter (see also Eq. (2.75))

$$^T\alpha_1 = \frac{^Tp(\text{RP I}) - {}^Tp(\text{RP II})}{1 - {}^Tp(\text{RP I})} \tag{3.33}$$

was calculated in two different approximations. The hfi in the radical PhCH$_2$ was always described semiclassically, while the spin dynamics in the radical PhCH$_2{}^{13}$CO was also described semiclassically within one approach, but within another approach was described in the proper quantum-mechanical way. Thus, $^Tp(\text{RP I})$ was calculated using Eq. (3.30) in both approaches. Within one approach, $^Tp(\text{RP II})$ was also calculated using the same Eq. (3.30). Within a theoretically more consistent approach eight algebraic linear equations, mentioned above, were numerically calculated to find $^Tp(\text{RP II})$. The results of these calculations are presented in Fig. 3.15.

These two curves show that both approaches yield very similar results. Thus, for RPs with many magnetically non-equivalent nuclei, the MIE parameters can be calculated with reasonable accuracy in the framework of the semiclassical description of the hfi even when the scales of hf coupling constants differ substantially.

3.4 Magnetic isotope effect induced by anisotropic hyperfine interaction (paramagnetic relaxation)

Paramagnetic relaxation induced by an anisotropic part of the hfi becomes important when, e.g., a reaction proceeds in a micelle [43]. In zero magnetic field the kinetic equations describing singlet-triplet transitions via the paramagnetic relaxation are

$$\frac{\partial n_S}{\partial t} = -3W n_S + W n_T \ ,$$

$$\frac{\partial n_T}{\partial t} = 3W n_S - W n_T \ . \tag{3.34}$$

In these equations n_S and n_T are the populations of the singlet and triplet states, respectively, and the relaxation rate W is given by (see also Eq. (2.56))

$$W = \sum_k \frac{2}{3} I_k (I_k + 1)(\gamma_e \, \gamma_{nk} \hbar)^2 r_k^{-6} \tau_0 \ , \tag{3.35}$$

where the summation includes the contribution to the relaxation rate from the hfi with all RP magnetic nuclei. The RP recombination probability with consideration of S-T transitions induced by paramagnetic relaxation was found in [34] for the continuous diffusion and the two-position RP models. For triplet-born RPs the results are:

$$^T p(S) = \frac{K_S \tau_r f}{4 \varDelta} \ ,$$

$$\varDelta = 1 + K_S \tau_r + f \left(1 + \frac{K_S \tau_r}{4} \right) \ , \tag{3.36}$$

where within the continuous diffusion model

$$f = \sqrt{4 W \tau_D} \tag{3.37}$$

and within the two-position model (see Eqs. (2.79)–(2.86))

$$f = \frac{\tau}{\tau_2} \frac{4 W \tau_m}{1 + 4 W \tau_m} \ . \tag{3.38}$$

At the isotope substitution, the relaxation rate changes proportionally to the square of the nuclear magnetic moments. For instance, at the H $\rightarrow$ D

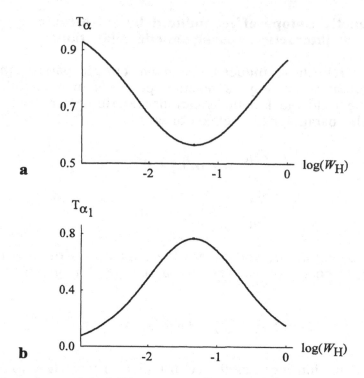

Fig. 3.16. Dependence of the MIE parameters $^T\alpha$ (**a**) and $^T\alpha_1$ (**b**) on the relaxation rate W_H (Eqs. (3.12) and (3.13)) at the H → D isotope substitution (W_H in ns^{-1} units). The parameters used for these calculations are: $K_S = 1000$ ns^{-1}, $\tau = 100$ ns, $\tau_2 = 5$ ns, and $\tau_r = 0.01$ ns.

isotope substitution the nuclear magnetic moment decreases four times. Thus, MIE may be quite pronounced in the case of long-lived RPs. Figure 3.16 illustrates the dependence of the parameters $^T\alpha$ and $^T\alpha_1$ (Eqs. (3.12) and (3.13)) at the H → D isotope substitution as a function of the relaxation rate in protonated RPs. In this figure W_H changes from 10^6 to 10^9 s^{-1}. When W_H increases, MIE disappears since for both isotope compositions hfi provides the efficient S-T mixing. MIE also disappears when the relaxation process slows down. For the particular set of parameters presented in Fig. 3.16, $5 \cdot 10^7$ s^{-1} is an optimal value of W_H.

3.5 Magnetic isotope effect in reaction kinetics

The previous discussion of a magnetic isotope substitution influence was connected with an isotope enrichment parameter which reflects the variation of the RP recombination probability at isotope substitution. However, an isotope substitution can affect the kinetic behaviour of the RP recombination as well. This is rather evident from the kinetic scheme of an RP. For instance, the two-position model of an RP is described by the following scheme (see also Fig. 2.10)

$$M \xleftarrow{(K)} {}^S\{R_1 \cdot \ \cdot R_2\}_1 \underset{1/\tau_2}{\overset{1/\tau_1}{\rightleftarrows}} {}^S\{R_1 \cdot \ \cdot R_2\}_2 \xrightarrow{1/\tau} R_1, R_2$$

$$\updownarrow \quad \text{(S-T transitions)}$$

$$ {}^T\{R_1 \cdot \ \cdot R_2\}_1 \underset{1/\tau_2}{\overset{1/\tau_1}{\rightleftarrows}} {}^T\{R_1 \cdot \ \cdot R_2\}_2 \xrightarrow{1/\tau} R_1, R_2 \ . \quad (3.39)$$

According to this scheme S-T transitions in RPs are of critical importance when discussing kinetic features such as time dependence of the RP population in a singlet or triplet state (n_S and n_T, respectively), or time dependence of the total amount of RPs ($n = n_S + n_T$). Using a quasistationary approximation for the RPs in the reaction zone (see Eqs. (2.81)) we have the following kinetic equations in the case of S-T transitions induced by paramagnetic relaxation (see Eqs. (3.34))

$$\frac{\partial n_S}{\partial t} = -3W n_S + W n_T - \frac{1}{\tau_2} \frac{K_S \tau_1}{1 + K_S \tau_1} n_S - \frac{1}{\tau} n_S \ ,$$

$$\frac{\partial n_T}{\partial t} = 3W n_S - W n_T - \frac{1}{\tau_2} n_T - \frac{1}{\tau} n_T \ . \quad (3.40)$$

The solution of these equations, e.g., for triplet-born RPs yields two-exponential time dependence of RP population in singlet and triplet state:

$$n_S = \frac{W}{2V} \left[\exp(-K_2 t) - \exp(-K_1 t) \right] \ ,$$

$$n_T = \exp(-K_2 t) + \frac{1}{2V} \left(V - W - \frac{\lambda}{2\tau_2} \right)$$

$$\times \left[\exp(-K_2 t) - \exp(-K_1 t) \right] \ . \quad (3.41)$$

where the rates K_1 and K_2 are equal to

$$K_1 = 2W + \frac{1}{\tau} + \frac{\lambda}{2\tau_2} + V \ ,$$

$$K_2 = 2W + \frac{1}{\tau} + \frac{\lambda}{2\tau_2} - V \quad,$$

$$V = \sqrt{3W^2 + \left(W + \frac{\lambda}{2\tau_2}\right)^2} \quad,$$

$$\lambda = \frac{K_S \tau_1}{1 + K_S \tau_1} \quad. \tag{3.42}$$

Let us consider limiting situations of fast and slow S-T mixing (see also the discussion of this problem within the one-position RP model in [43]). At the fast paramagnetic relaxation, when $W > \lambda / 2\tau_2$,

$$K_1 = 4W + \frac{1}{\tau} + \frac{3\lambda}{4\tau_2} \quad,$$

$$K_2 = \frac{1}{\tau} + \frac{\lambda}{4\tau_2} \quad. \tag{3.43}$$

and at the relatively slow paramagnetic relaxation, when $W < \lambda / 2\tau_2$,

$$K_1 = 3W + \frac{1}{\tau} + \frac{\lambda}{\tau_2} \quad,$$

$$K_2 = W + \frac{1}{\tau} \quad. \tag{3.44}$$

Equations (3.44) demonstrate that S-T transitions induced by paramagnetic relaxation can contribute to both decay components, the fast one with K_1, and the slow one with K_2. Figures 3.17 and 3.18 illustrate the kinetic behaviour of the RP populations for the two sets of kinetic parameters. These curves may correspond to the recombination of RPs with two different isotope compositions. One can see that all RP populations (singlet state, triplet state and total RP population) are affected by S-T transitions. The triplet RP population shows that two ranges of decay – fast and slow – exist. These curves give one the idea that it would be most promising to detect the magnetic isotope effect on the initial range of the singlet and/or triplet RP population variation. It was shown [44, 45] that in the framework of the continuous diffusion model of RPs similar regularities operate.

Similarly, the isotropic hyperfine coupling can affect the kinetics of the RP recombination. With the help of Eqs. (2.81), the kinetics of the RP recombination was calculated taking into account S-T transitions induced by the isotropic hfi with one magnetic nucleus of spin $I = 1/2$. The results of these numerical calculations are depicted in Figs. 3.19 and 3.20.

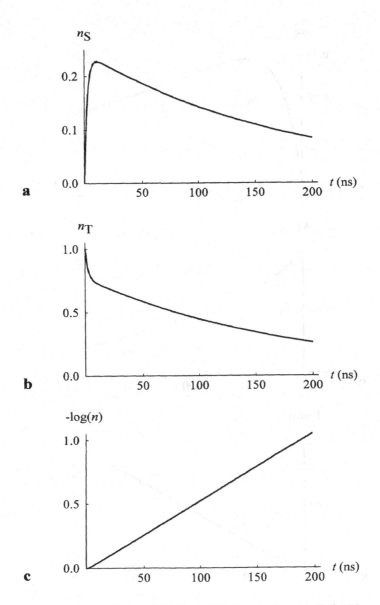

Fig. 3.17. Time dependence of the triplet-born RP decay characteristics describing a fraction of **a** the singlet (n_S) and **b** triplet (n_T) RPs, and **c** the total amount of RPs ($n = n_S + n_T$) calculated within the two-position model of RPs including singlet-triplet transitions induced by paramagnetic relaxation with $W = 0.1$ ns^{-1}. The parameters used for these calculations are: $\lambda = 0.9$, $\tau_2 = 50$ ns, and $\tau = 1000$ ns.

Two features of the results presented in Figs. 3.19 and 3.20 deserve to be noted. First, about one third of triplet-born RPs decay rather fast, the other two thirds of triplet-born RPs decay slowly. This reflects the fact that only one third of triplet-born RPs can be converted to the sin-

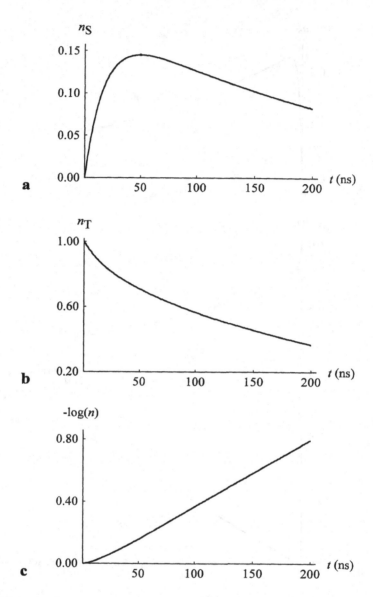

Fig. 3.18. Time dependence of the triplet-born RP decay characteristics describing a fraction of **a** the singlet (n_S) and **b** triplet (n_T) RPs, and **c** the total amount of RPs ($n = n_S + n_T$) calculated within the two-position model of RPs including singlet-triplet transitions induced by paramagnetic relaxation with $W = 0.01$ ns^{-1}. The parameters used for these calculations are: $\lambda = 0.9$, $\tau_2 = 50$ ns, and $\tau = 1000$ ns.

glet state and then recombine if the isotropic hf coupling with one magnetic nucleus is the main driving force for S-T mixing. For this particular situation two thirds of triplet-born RPs disappear within the lifetime τ of the two-position model (see, e.g., Eq. (3.39)). Secondly, the

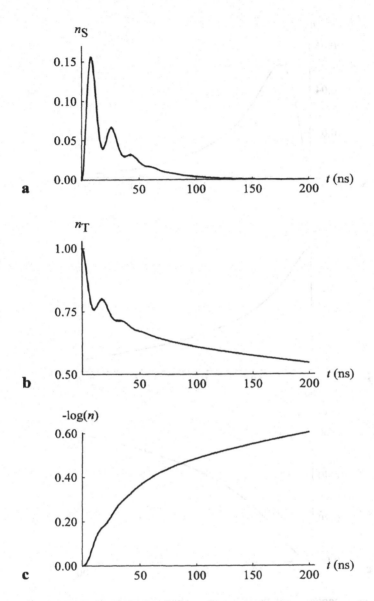

Fig. 3.19. Time dependence of the RP decay characteristics describing a fraction of **a** the singlet (n_S) and **b** triplet (n_T) RPs, and **c** the total amount of RPs ($n = n_S + n_T$). It is assumed that RP has only one magnetic nucleus. The hf coupling constant is 2 mT. Other parameters are: $\lambda = 0.9$, $\tau_2 = 10$ ns, and $\tau = 1000$ ns.

kinetic curves in Figs. 3.19 and 3.20 are not monotonously varying, they exhibit oscillations due to the oscillatory character of S-T dynamics induced by the isotropic hf coupling. The oscillations are not pronounced if the hf coupling constant is small enough (compare Figs. 3.19 and 3.20). It should be also noted that the total amount of RPs manifests

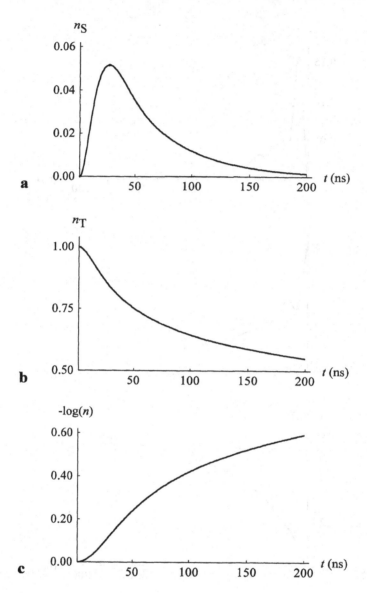

Fig. 3.20. Time dependence of the RP decay characteristics describing a fraction of **a** the singlet (n_S) and **b** triplet (n_T) RPs, and **c** the total amount of RPs ($n = n_S + n_T$). It is assumed that RP has only one magnetic nucleus. The hf coupling constant is 0.5 mT. Other parameters are: $\lambda = 0.9$, $\tau_2 = 10$ ns, and $\tau = 1000$ ns.

very little oscillations in contrast to the populations of the singlet and triplet states separately (see Fig. 3.19).

There are several sources which can smear out oscillations of the population of RPs in singlet and triplet state. This can arise from the simultaneous action of the isotropic hfi with many magnetically non-equivalent nuclei. However, even in RPs with only one or a few magnetic nuclei demonstrating the isotropic hfi, the oscillations under discussion can be smeared out as a result of a contribution of the paramagnetic relaxation to S-T mixing alongside with the isotropic hfi. To illustrate this behaviour, the recombination kinetics was calculated for the model RP in a situation where S-T transitions are caused by the combined action of the isotropic hf coupling with one nucleus of spin $I = 1/2$ and the paramagnetic relaxation. The results are shown in Figs. 3.21 and 3.22. These figures have to be compared with Fig. 3.19: the relaxation diminishes the oscillatory behaviour and accelerates the whole process of the RP disappearance.

A contribution of the hfi to S-T dynamics can be substantially reduced by decreasing the RP lifetime between subsequent re-encounters within the reaction regime. For instance, for the two-position model of RP kinetic curves were calculated for the recombination of RPs with $\tau_2 = 1$ ns and other parameters as in Fig. 3.19. The results are presented in Fig. 3.23.

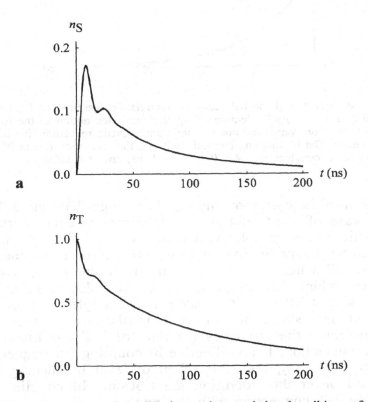

Fig. 3.21. Time dependence of the RP decay characteristics describing a fraction of **a** the singlet (n_S) and **b** triplet (n_T) RPs demonstrating the combined action of the isotropic hyperfine interaction with one particular nucleus and paramagnetic relaxation due to the anisotropic hfi with nuclei. The hf coupling constant is 2 mT, the relaxation rate is $W = 0.01$ ns^{-1}. Other parameters are: $\lambda = 0.9$, $\tau_2 = 10$ ns, and $\tau = 1000$ ns.

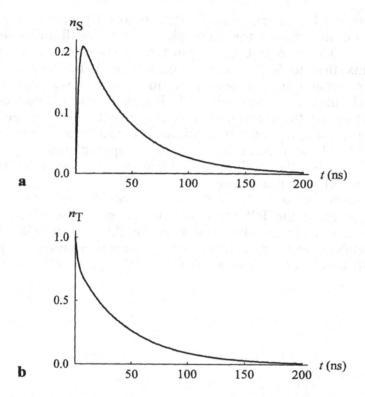

Fig. 3.22. Time dependence of the RP decay characteristics describing a fraction of **a** the singlet (n_S) and **b** triplet (n_T) RPs demonstrating the combined action of the isotropic hyperfine interaction with one particular nucleus and paramagnetic relaxation due to the anisotropic hfi with nuclei. The hf coupling constant is 2 mT, the relaxation rate is $W = 0.1 \text{ ns}^{-1}$. Other parameters are: $\lambda = 0.9$, $\tau_2 = 10$ ns, and $\tau = 1000$ ns.

Comparison of the corresponding curves in Figs. 3.19 and 3.23 shows that, in the case of short lifetime between two re-encounters of pair radicals, kinetic curves do not reveal quantum oscillations. Qualitatively, this result can be interpreted in terms of a broadening of an energy level of the singlet RP state, since at each encounter the singlet component of RPs can recombine, i.e., disappear. As a result, due to the uncertainty principle, the singlet RP level is broadened. The singlet state level broadening will not only smear out quantum oscillations of kinetic curves, but will also reduce the efficiency (or rate) of S-T transitions induced by hyperfine interaction. If the effective hf coupling (in frequency units rad/s) is less than λ/τ_2, then this effect of level broadening becomes important, and under this condition the isotropic hf coupling induces non-oscillating S-T conversion with the rate

$$K_{\text{ef}} = \frac{a_{\text{ef}}^2 \tau_2}{\lambda} . \qquad (3.45)$$

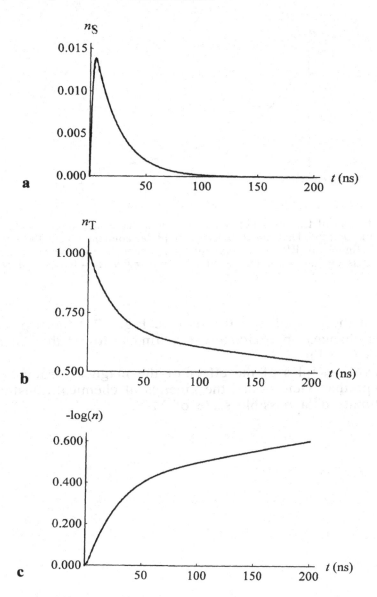

Fig. 3.23. Time dependence of the RP decay characteristics describing a fraction of **a** the singlet (n_S) and **b** triplet (n_T) RPs, and **c** the total amount of RPs ($n = n_S + n_T$) demonstrating the effect of the lifetime between two subsequent encounters of RP partners. The parameters used for these calculations are: $A = 2$ mT, $\lambda = 0.9$, $\tau = 1000$ ns, and $\tau_2 = 1$ ns.

To illustrate this feature of the RP recombination kinetics, in Fig. 3.24 we present the kinetic curves for the recombination of RPs with $\tau_2 = 0.5$ ns. All other parameters are the same as in Fig. 3.23.

Comparison of kinetic curves for two different values of τ_2 (Figs. 3.23 and 3.24) confirms our qualitative prediction. In accordance with

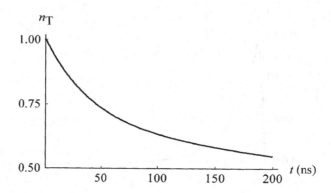

Fig. 3.24. Reduction of triplet-singlet conversion with decreasing of the lifetime between re-encounters of radicals. This kinetic curve should be compared with the curve in Fig. 3.23b. For both figures all RP parameters coincide, except the time between re-encounters (in this figure τ_2 is two times less than in Fig. 3.23): $\lambda = 0.9$, $\tau_2 = 0.5$ ns, and $\tau = 1000$ ns.

Eq. (3.45), in the case of $\tau_2 = 0.5$ ns (see Fig. 3.24) the triplet population of RPs changes approximately two times slower than in the case of $\tau_2 = 1$ ns (see Fig. 3.23).

The above examples of the effect of the magnetic isotope substitution on the product yield or/and the kinetics of chemical transformations give an estimate of a possible scale of MIE.

4 Magnetic isotope effect in the presence of external magnetic fields

In this chapter we will discuss one of the MIE's most remarkable features: its dependence on constant and alternating external magnetic fields. The influence of the external magnetic fields on the isotope effect parameters suggests that MIE operates. The physical origin of the phenomenon was briefly discussed in Chap. 2. The constant field splits the triplet sublevels (see, e.g., Fig. 2.6) so that the transitions between the RP singlet S and T_{+1}, T_{-1} triplet sublevels induced by the hyperfine interaction with the magnetic nuclei can be suppressed at high field intensities. The external field also affects spin conservation properties (see the discussion of this problem in Sect. 2.6). Alternating fields are able to affect spin dynamics in the RPs with a definite isotope composition as well. How an external magnetic field influences the contribution of the hfi to the RP recombination depends on the spin-spin interactions (Heisenberg exchange and dipole-dipole interaction) between unpaired electrons. For that reason this chapter also includes a brief discussion of the role played by the interradical interaction in the RP spin dynamics. This problem may be of great significance for reactions that proceed through biradical states or for reactions in restricted spaces like micelles. At high constant magnetic field intensities, the mechanisms of RP singlet-triplet transitions caused by the difference in Larmor frequencies of two radicals in a pair (the so-called Δg-mechanism) and by paramagnetic relaxation originating from anisotropy of the radicals g-tensors start operating. These mechanisms of RP S-T transitions which are not sensitive to the isotope composition reduce the role of the hfi induced S-T transitions and, as a result, the scale of the MIE parameters at high magnetic fields decreases as well. Therefore, this chapter includes also a brief discussion of the mechanisms of S-T transitions which compete with hfi.

4.1 Magnetic isotope effect as affected by constant fields

The magnetic isotope effect reflects a variation in the contribution of hfi induced S-T transitions to the RP recombination at isotope substitution. To specify the situation, throughout this section we assume that

RPs have triplet precursor molecules and that they are able to recombine only from their singlet state. For such a situation the RP recombination probability increases with an increase of S-T mixing efficiency, since the intersystem S-T crossing pumps RPs from the non-reactive triplet state to the reactive singlet state.

4.1.1 Anisotropic hyperfine interaction (the so-called relaxation mechanism of singlet-triplet transitions)

We start with discussing the following model situation. Assume that RPs have triplet precursors, while they can recombine only from their singlet state. Further we assume that interradical spin-spin interactions (Heisenberg exchange and dipole-dipole) can be neglected when two partners of a radical pair are well separated, i.e., in time intervals between two subsequent re-encounters of radicals. Under these circumstances, if the anisotropic hfi is the predominant contributor to S-T transitions, then the RP recombination probability has to decrease with an increase in external field intensity. The anisotropic hfi is modulated by the random rotational motion of radicals, and its contribution to the spin evolution of the RP unpaired electrons at least in non-viscous solvents can be described in terms of the paramagnetic relaxation times T_1 and T_2. According to Eqs. (2.56) the rates of the hfi induced paramagnetic relaxation diminish with an increase in the field intensity B_0. The kinetic equations describing the time variation of the population of RPs in singlet and triplet states, caused by the paramagnetic relaxation, are [5, 34, 45, 46]

$$\left(\frac{\partial \rho}{\partial t}\right)_{SS} = -\left(\frac{1}{4T_1'} + \frac{1}{2T_2'}\right)\rho_{SS} + \left(-\frac{1}{4T_1'} + \frac{1}{2T_2'}\right)\rho_{T_0T_0}$$
$$+ \frac{1}{4T_1'}\left(\rho_{T_{+1}T_{+1}} + \rho_{T_{-1}T_{-1}}\right),$$

$$\left(\frac{\partial \rho}{\partial t}\right)_{T_0T_0} = \left(-\frac{1}{4T_1'} + \frac{1}{2T_2'}\right)\rho_{SS} - \left(\frac{1}{4T_1'} + \frac{1}{2T_2'}\right)\rho_{T_0T_0}$$
$$+ \frac{1}{4T_1'}\left(\rho_{T_{+1}T_{+1}} + \rho_{T_{-1}T_{-1}}\right),$$

$$\left(\frac{\partial \rho}{\partial t}\right)_{T_{+1}T_{+1}} = \frac{1}{4T_1'}\left(\rho_{SS} + \rho_{T_0T_0}\right) - \frac{1}{2T_1'}\rho_{T_{+1}T_{+1}} - \frac{1}{2T_1''}\,\mathrm{Re}\rho_{ST_0},$$

$$\left(\frac{\partial \rho}{\partial t}\right)_{T_{-1}T_{-1}} = \frac{1}{4T_1'}\left(\rho_{SS} + \rho_{T_0T_0}\right) - \frac{1}{2T_1'}\rho_{T_{-1}T_{-1}} + \frac{1}{2T_1''}\,\mathrm{Re}\rho_{ST_0},$$

$$\left(\frac{\partial \rho}{\partial t}\right)_{ST_0} = -\left(\frac{1}{4T_1'} + \frac{1}{2T_2'}\right)\rho_{ST_0} + \left(-\frac{1}{4T_1'} + \frac{1}{2T_2'}\right)\rho_{T_0S}$$
$$- \frac{1}{4T_1''}\left(\rho_{T_{+1}T_{+1}} - \rho_{T_{-1}T_{-1}}\right), \tag{4.1}$$

where

$$\frac{1}{T_1'} = \frac{1}{T_{1A}} + \frac{1}{T_{1B}}, \qquad \frac{1}{T_2'} = \frac{1}{T_{2A}} + \frac{1}{T_{2B}}, \qquad \frac{1}{T_1''} = \frac{1}{T_{1A}} - \frac{1}{T_{1B}},$$

where $1/T_{1A}$, $1/T_{1B}$, $1/T_{2A}$, $1/T_{2B}$ are the rates of the longitudinal ($1/T_1$) and the transverse ($1/T_2$) relaxation of radicals A and B of a pair, respectively. The contribution of each magnetic nucleus to the paramagnetic relaxation rates is given by the following expressions (see also Eqs. (2.56))

$$\frac{1}{T_1} = \frac{2W}{1 + (\gamma_e B_0 \tau_0)^2},$$
$$\frac{1}{T_2} = W\left(1 + \frac{1}{1 + (\gamma_e B_0 \tau_0)^2}\right), \tag{4.2}$$

where $W = (2/3)I(I + 1)(\gamma_e \gamma_n \hbar)^2 r^{-6} \tau_0$, r is the distance between the electron and the nucleus, τ_0 is the correlation time of the radical's rotational motion, and B_0 is the intensity of the external magnetic field.

In the presence of many magnetic nuclei, their contributions to the relaxation rates are added (see, e.g., Eq. (3.33)). From Eqs. (4.2) it follows that $1/T_1$ tends towards zero at high magnetic field intensities. Thus, in the high field limit the kinetic Eqs. (4.1) reduce to

$$\left(\frac{\partial \rho}{\partial t}\right)_{SS} = -W\rho_{SS} + W\rho_{T_0T_0},$$

$$\left(\frac{\partial \rho}{\partial t}\right)_{T_0T_0} = W\rho_{SS} - W\rho_{T_0T_0},$$

$$\left(\frac{\partial \rho}{\partial t}\right)_{T_{+1}T_{+1}} = 0,$$

$$\left(\frac{\partial \rho}{\partial t}\right)_{T_{-1}T_{-1}} = 0,$$

$$\left(\frac{\partial \rho}{\partial t}\right)_{ST_0} = -\frac{1}{2T_2'}\rho_{ST_0} + \frac{1}{2T_2'}\rho_{T_0S}. \tag{4.3}$$

These equations demonstrate that at high fields the hfi induced para-magnetic relaxation connects only one triplet sublevel, namely the T_0 state, with the singlet state. Thus, with an increase in field intensity the efficiency of S-T mixing decreases, and the number of the S-T conver-sion channels decreases from three to one. Simultaneously, the rate of S-T_0 mixing decreases two times with an increase in field intensity. The relaxation rates decrease sharply when the Larmor frequency of unpaired electrons becomes higher than the characteristic frequency of molecular rotational diffusion, $1/\tau_0$, (see Eq. (4.2)). Thus this transition occurs around the field value

$$B_0^* \approx \frac{1}{\gamma_e \tau_0} . \tag{4.4}$$

For $\tau_0 \approx 10^{-10}$–10^{-11} s this transition range will be at $B_0^* \approx 0.05$–0.5 T, i.e., at external magnetic field intensities which are much higher than the local hfi fields. The field dependence of the RP recombination prob-ability, when calculated on the basis of Eqs. (2.87), (4.1) and (4.2) as

$$^{\mathrm{T}}p_{\mathrm{g}}(\mathrm{S}) = \frac{\lambda}{\tau_2} \int_0^\infty \rho_{\mathrm{SS}}(t)\, \mathrm{d}t , \tag{4.5}$$

is presented in Fig. 4.1.

The curves p_{H} and p_{D} give the recombination probability of the pro-tonated and deuterated RPs, respectively. Two features of the curves in Fig. 4.1 are to be mentioned. Figure 4.1 shows that RP recombination probabilities decrease with increasing field intensity. The curves also show that p_{H} decreases at somewhat higher values of the field intensity than happens in a case of p_{D}. This observation can be interpreted as follows. S-T transitions operate in a competition with other processes in RPs. Therefore, the RP recombination probability changes when S-T transitions become less efficient as compared to rates of other processes in RPs. The slower the relaxation rates at zero external field, the lower external field intensities suffice to diminish the contribution of S-T tran-sitions during the RP lifetime.

Figure 4.2 shows the MIE parameters for the situation presented in Fig. 4.1.

From Fig. 4.2 we see that the field dependence of the MIE param-eters may demonstrate non-monotonous behaviour. Figure 4.2 indicates that $^{\mathrm{T}}\alpha_1$ can pass through the extreme at an intermediate field intensity. However, $^{\mathrm{T}}\alpha$ can also exhibit an extreme. For instance, Fig. 4.3 displays the same MIE parameters as in Fig. 4.2 but for another set of the RP parameters. Here we see that $^{\mathrm{T}}\alpha$ also passes through the extreme.

Figures 4.2 and 4.3 demonstrate that the MIE parameters can reveal non-monotonous field dependence even when the RP recombination

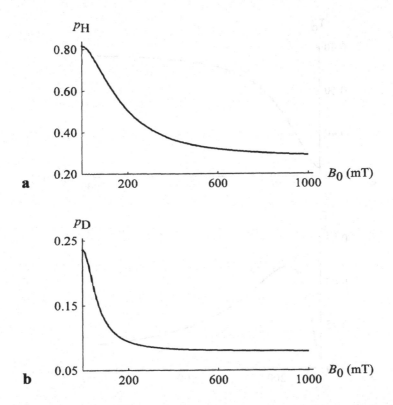

Fig. 4.1. Dependence of the triplet-born RP recombination probability on the field B_0 calculated within the two-position model. The curves p_H and p_D differ in the values of the paramagnetic relaxation parameter W: $W_H = 5 \cdot 10^6\,\text{s}^{-1}$ in case of p_H (**a**) and W_D is 16 times less in case of p_D (**b**). Other parameters used for these calculations are: $\lambda = 0.9$, $\tau = 1000$ ns, $\tau_2 = 5$ ns, and $\tau_0 = 0.1$ ns.

probabilities for each isotope composition of radicals have monotonous variation with increasing field intensity.

However, in the presence of an external field, the difference of the Zeeman frequencies of the radicals and the paramagnetic relaxation induced by the random modulation of the anisotropic part of the Zeeman interaction (anisotropic g-tensor) may contribute essentially to S-T dynamics in RPs (see discussion of this subject in [5, 34, 45, 46]). Being independent of the isotope composition, the contribution of the Zeeman interaction to S-T transitions in RPs masks the effect of the hfi and, as a result, diminishes the magnetic isotope effect. To illustrate this behaviour, the MIE parameters were calculated at H → D isotope substitution for the same model situation as for Figs. 4.1 and 4.2 with only one exception: it was assumed that the isotropic g-factors of the two radicals were different, $\Delta g = 0.002$. The results of these calculations are shown in Fig. 4.4.

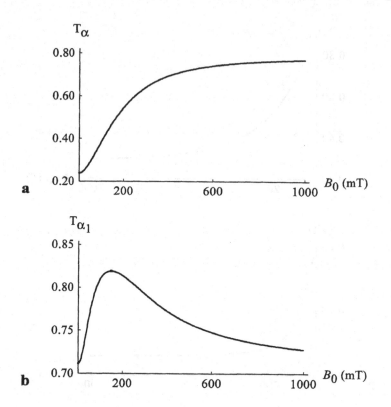

Fig. 4.2. Dependence of the MIE parameters $^T\alpha$ (**a**) and $^T\alpha_1$ (**b**) on the field B_0 defined by Eqs. (3.12) and (3.13) at the H $\rightarrow$ D isotope substitution in triplet-born RP recombination. The parameters used for these calculations are: $W_H = 5 \cdot 10^6$ s^{-1}, $W_D = W_H/16$, $\lambda = 0.9$, $\tau = 1000$ ns, $\tau_2 = 5$ ns, and $\tau_0 = 0.1$ ns.

The rate of S-T transitions induced by the Zeeman frequency difference linearly increases with the field intensity. The corresponding transition matrix element equals (see Eqs. (2.41)):

$$H_{ST_0} = \frac{1}{2}(g_A - g_B)\beta B_0 \ . \tag{4.6}$$

This leads to an increase of triplet-born RP recombination probabilities at high fields (see curves p_H and p_D in Fig. 4.4). Simultaneously, the magnetic isotope effect disappears in high magnetic fields. Note that the MIE parameters $^T\alpha = 1$ and $^T\alpha_1 = 0$ describe the situation without any isotope effect (see Eqs. (3.12) and (3.13)).

The contribution of the hfi to the RP recombination kinetics also depends on the magnetic field strength. Thus, the field dependence of reaction rates at isotope substitution allows for some freedom in studying the MIE phenomenon and in the applications of this effect in basic

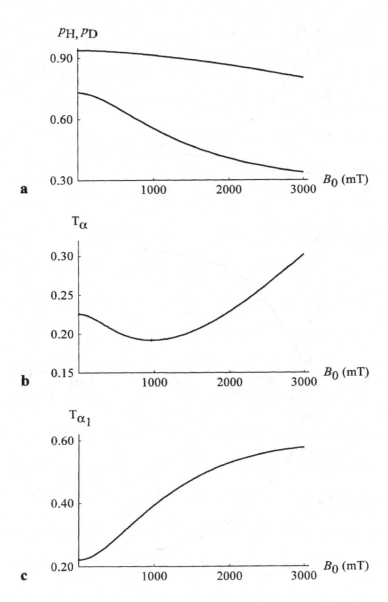

Fig. 4.3. Dependence of the triplet-born RP recombination probabilities (**a**; p_H, top curve; p_D, bottom curve) and of the MIE parameters $^T\alpha$ (**b**) and $^T\alpha_1$ (**c**) on the field B_0 during H $\rightarrow$ D isotope substitution calculated within the two-position model. These curves are to be compared correspondingly with those presented in Figs. 4.1 and 4.2. The parameters used for these calculations are: $W_H = 5 \cdot 10^7$ s^{-1}, $W_D = W_H/16$, $\lambda = 0.9$, $\tau = 1000$ ns, $\tau_2 = 10$ ns, and $\tau_0 = 0.01$ ns.

researches and technology. In low fields, the relaxation transitions are able to convert RPs from all triplet sublevels to the singlet state maintaining practically the same rates. In high fields, however, transition rates

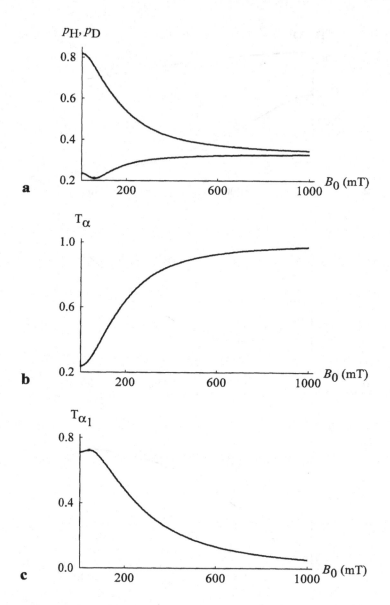

Fig. 4.4. Dependence of the RP recombination probabilities (**a**; p_H, top curve; p_D, bottom curve) and of the MIE parameters $^T\alpha$ (**b**) and $^T\alpha_1$ (**c**) on the field B_0 demonstrating the reduction of MIE by the Δg-mechanism of S-T transitions. These curves are to be compared correspondingly with those presented in Figs. 4.1 and 4.2. The parameters used for these calculations are: $W_H = 5 \cdot 10^6$ s^{-1}, $W_D = W_H/16$, $\Delta g = 0.002$, $\lambda = 0.9$, $\tau = 1000$ ns, $\tau_2 = 5$ ns, and $\tau_0 = 0.1$ ns.

from two of the triplet sublevels, namely from the T_{+1} and T_{-1} states decrease remarkably. Thus, the two-exponential behaviour of an RP decay should be more pronounced in high magnetic fields than in low

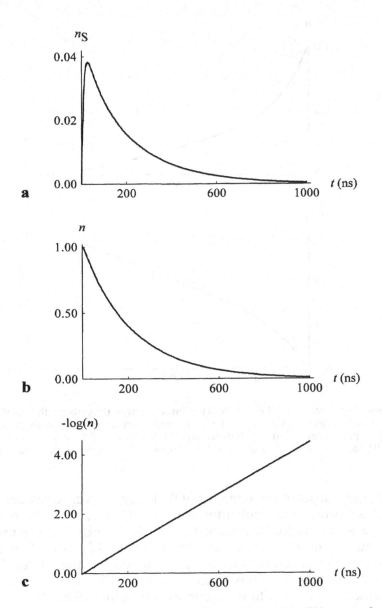

Fig. 4.5. Time dependence of the RP decay characteristics describing **a** a fraction of the singlet (n_S) RPs and **b** and **c** the total amount of RPs ($n = n_S + n_T$). The parameters used for these calculations are: $\lambda = 0.9$, $\tau = 1000$ ns, $\tau_2 = 10$ ns, $\tau_0 = 0.01$ ns, $W = 10^7\,\mathrm{s}^{-1}$, and $B_0 = 0.1$ T. Triplet RP precursors are considered.

fields. This feature of RP recombination kinetics is illustrated by Figs. 4.5 and 4.6. Indeed, RP decay can be perfectly described by only one exponent (Fig. 4.5) for relatively low magnetic fields, e.g., $B_0 = 0.1$ T, while at $B_0 = 1$ T (Fig. 4.6) the kinetic curves definitely show the existence of fast and slow RP decay.

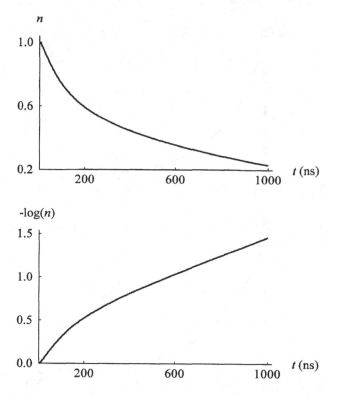

Fig. 4.6. Time dependence of the RP decay characteristics describing the total amount of RPs ($n = n_S + n_T$). These curves are to be compared correspondingly with those presented in Fig. 4.5. The parameters used for these calculations are: $\lambda = 0.9$, $\tau = 1000$ ns, $\tau_2 = 10$ ns, $\tau_0 = 0.01$ ns, $W = 10^7$ s^{-1}, and $B_0 = 1$ T. Triplet RP precursors are considered.

An isotope substitution changes RP decay curves considerably. For instance, if all protons are substituted by deuterons, then the hfi induced relaxation rates are reduced by the factor 16. In Fig. 4.7 we present the RP decay curve for the same parameters as in Fig. 4.6. The only difference is that the relaxation rates are reduced by the factor 16.

The curves in Fig. 4.7 show that with $W = 10^7/16$ s^{-1} the role of relaxation induced S-T transitions becomes negligible. RPs decay with their total lifetime $\tau = 1000$ ns.

4.1.2 Isotropic hyperfine interaction

The external magnetic field dependence of the RP recombination and of the MIE parameters has several peculiar features when S-T transitions in RPs are governed by the isotropic hf coupling. Let us consider the recombination of triplet-born RPs. In Sect. 2.6, it was shown that the recombination probability of triplet-born RPs can pass through a

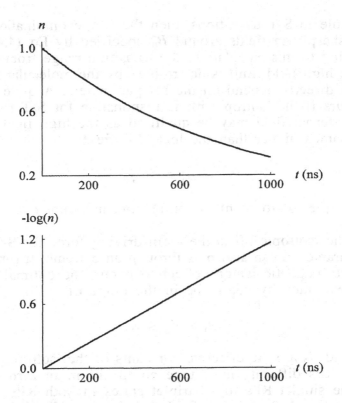

Fig. 4.7. Time dependence of the RP decay characteristics describing the total amount of RPs ($n = n_S + n_T$). These curves are to be compared correspondingly with those presented in Fig. 4.6. The parameters used for these calculations are: $\lambda = 0.9$, $\tau = 1000$ ns, $\tau_2 = 10$ ns, $\tau_0 = 0.01$ ns, $B_0 = 1$ T, and $W = 10^7/16$ s^{-1}. Triplet RP precursors are considered.

maximum for radicals with magnetically equivalent nuclei or for radicals with only a few magnetic nuclei. This extreme arises from the fact that in zero magnetic field, the isotropic hfi does not change the total electron and nuclear spin of RPs [39]. In the presence of an external field, this spin conservation rule is violated and the efficiency of S-T mixing induced by the isotropic hfi increases with an increase in field strength. However, S-T mixing caused by the isotropic hfi decreases in high fields since, in high fields, the isotropic hfi can only mix the singlet state with one triplet substate T_0. Other two triplet sublevels T_{+1} and T_{-1} are out of resonance with the singlet state (see Fig. 2.6). If RPs possess many magnetically non-equivalent nuclei, the RP recombination probability decreases monotonously with increasing field intensity. Therefore, in a case of RPs with many magnetically non-equivalent nuclei the field dependence of the RP recombination probability resembles that when paramagnetic relaxation induced by the anisotropic hfi operates. However, there is also an important difference between these two situations. As we have seen above, if the paramagnetic relaxation plays a dominant role in S-T transitions, then the RP recombination probability decreases sharply in fields around B_0^* specified by Eq. (4.4).

According to this calculation, the transition range from the low field limit to the high field limit is determined by the molecular mobility (τ_0). It does not directly depend on the hfi parameters. A quite different situation occurs if the isotropic hfi is responsible for S-T mixing. In this case, the external field may be qualified as the high field when its intensity becomes higher than the local hfi field

$$B_0^* > A \; , \tag{4.7}$$

where A is the isotropic hf coupling constant quoted in the magnetic field units.

When the isotropic hfi is the main driving force for S-T transitions, the MIE parameters can also pass through an extreme. It can be expected that, for the magnetic isotope effect to occur, the optimal value of the external field intensity has to be in the range of

$$A_1 \leq B_0 \leq A_2 \; , \tag{4.8}$$

where A_1 and A_2 are the effective constants of the isotropic hf couplings in the two RPs differing in isotope composition. In zero field, the hfi converts the singlet RPs to all triplet states in both RPs. In the range specified by Eq. (4.8), the high field situation is realized for the RP with the weaker hf coupling (A_1): indeed, when $B_0 > A_1$ then only S-T$_0$ mixing is efficient for RPs with the hf coupling constant A_1, the two other channels S-T$_{+1}$, S-T$_{-1}$ hardly contribute at all to S-T mixing. However, for the RP with the stronger hf coupling (with hf coupling constant A_2), all three channels of the S-T conversion operate. When the external field strength becomes bigger than A_2, the high field situation occurs for both subensembles of RPs. Thus, it is justifiable to expect that the recombination probabilities of these two RPs will deviate noticeably in a field range specified by Eq. (4.8).

To illustrate the above points, the field dependence of the RP recombination probability was calculated for a model RP with one magnetic nucleus. The RP spin Hamiltonian contains the Zeeman interaction of electron spins S_A and S_B with the external field, the isotropic hfi and the exchange interaction between two radicals

$$\mathscr{H} = g_A \beta B_0 S_{Az} + g_B \beta B_0 S_{Bz} + \hbar a \, \mathbf{S_A} \cdot \mathbf{I} - \hbar \gamma_e J \left(\frac{1}{2} + 2 \mathbf{S_A} \cdot \mathbf{S_B} \right) \; , \tag{4.9}$$

where J is the exchange integral expressed in units of magnetic field induction. The numerical calculations were performed in the framework of the two-position model of RP. In this case, the kinetic equations are given by Eqs. (2.87) whereas the spin dynamics is determined by Eqs. (2.43) with the spin Hamiltonian (4.9). The results are presented in Fig. 4.8.

Figure 4.8 shows that the RP recombination probability passes through the maximum with an increase in the external field intensity. This observation coincides with our qualitative expectation. However, quantitatively, the inequality (4.7) gives a rather crude description of the field dependence of the RP recombination. The reason for this is that the RP recombination does not only depend on the parameters of the spin Hamiltonian. The total RP lifetime τ and the mean time between two subsequent re-encounters τ_2 influence the field dependence of the RP recombination probability as well. For example, Fig. 4.9 shows that the field dependence varies remarkably with the shortening of the mean time τ_2.

From Fig. 4.9 we see that the shorter the time interval τ_2, the broader the curve $p(B_0)$. This behaviour can be attributed to the broadening of the RP energy levels in accordance with the uncertainty principle: $\Delta E \approx \hbar/\tau_2$. Due to this energy level broadening effect, the singlet state overlaps with T_{+1} and T_{-1} states if the Zeeman splitting of T_{+1} and T_{-1} states does not exceed the level broadening, i.e., if $g\beta\hbar^{-1}B_0 \lesssim 1/\tau_2$.

In high magnetic fields, the difference of the Larmor frequencies of the two free radicals (the so-called Δg-mechanism) and/or the paramagnetic relaxation, caused by a random modulation of the anisotropic part of the Zeeman interaction, may contribute to S-T mixing. Their contribution will eliminate the difference in S-T mixing for RPs with different isotope composition. Therefore, in high magnetic fields, MIE should disappear if the unpaired electrons of the two radicals, which participate in the recombination, have different isotropic values of their g-factors or the RP unpaired electrons possess the anisotropic g-tensors. A

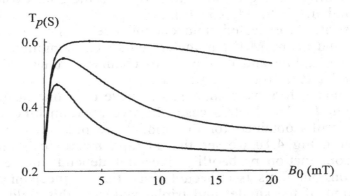

Fig. 4.8. Dependence of the triplet-born RP recombination probability on the field B_0. The curves correspond to the following values of the hf coupling constant A: 10 mT (top curve), 2 mT (middle curve), and 1 mT (bottom curve). Other parameters used for these calculations are: $\lambda = 0.9$, $\tau_2 = 5$ ns, $\tau = 200$ ns, $J = 0$, and $I = 1/2$.

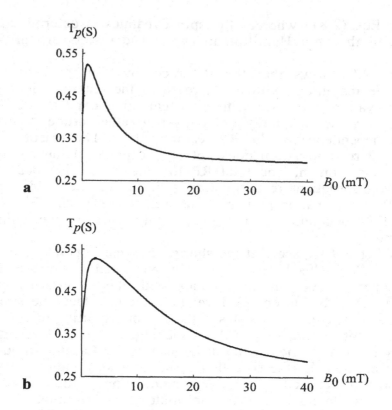

Fig. 4.9. Dependence of the RP recombination probability on the field B_0 for two values of τ_2: $\tau_2 = 10$ ns (**a**) and $\tau_2 = 1$ ns (**b**). Other parameters used for these calculations are: $A = 2$ mT, $\lambda = 0.9$, $\tau = 200$ ns, $J = 0$, and $I = 1/2$.

discussion of the contribution of Δg-mechanism and paramagnetic relaxation, originating from g-tensor anisotropy, to the spin evolution in RPs can be found in [5, 9, 34, 45, 46].

To illustrate an expected field dependence of the RP recombination probability and of the MIE parameters under the combined action of the isotropic hf coupling and of the Δg-mechanism, numerical calculations for two model RPs were undertaken.

In one model, RPs were supposed to have only one magnetic nucleus with the spin $I = 1/2$. In this case, the two-position kinetic model was used. The results obtained for this model are presented in Fig. 4.10.

First, from Fig. 4.10 we see that at high magnetic fields ($B_0 > 2$ T) the RP recombination probability does not depend on the isotropic hf coupling constant. This is expected since the Δg-mechanism dominates in the mixing of the singlet and triplet states in this field range. However, the curves in Fig. 4.10 manifest that there is a pronounced minimum. This minimum occurs when the field intensity matches the condition

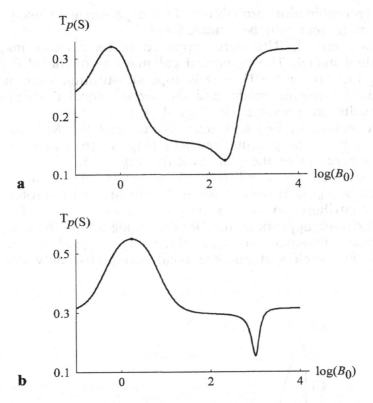

Fig. 4.10. Dependence of the triplet-born RP recombination probability on the field B_0 for one nucleus with spin $I = 1/2$ (B_0 in mT units). The curves correspond to the following values of the isotropic hf coupling constant A: 0.5 mT (**a**) and 2 mT (**b**). Other parameters used for these calculations are: $\lambda = 0.9$, $\tau_2 = 5$ ns, $\tau = 200$ ns, $J = 0$, and $\Delta g = 0.002$.

$$|\Delta g|\beta B_0 = \frac{1}{2}\hbar|a| . \qquad (4.10)$$

The physical origin of this minimum is as follows. In high magnetic fields, the ensemble of RPs with one nucleus of spin $I = 1/2$ can be divided into two subensembles of RPs with the projections $+1/2$ and $-1/2$ of the nuclear spin onto the external field direction. For the spin Hamiltonian (4.9), the S-T_0 transition matrix elements responsible for S-T mixing in RPs in high magnetic fields are equal to $(\Delta g\beta B_0 + \hbar a/2)/2$ and $(\Delta g\beta B_0 - \hbar a/2)/2$ for the RP subensembles with $+1/2$ and $-1/2$ nuclear spin projection, respectively. Thus, if the field intensity fits Eq. (4.10), the S-T_0 transition matrix element in one of the subensembles of RPs vanishes. This particular subensemble of RPs will not convert from the initial triplet state to the reactive singlet state under the condition (4.10). As a result, the minimum occurs in the field dependence of the RP recombination probability. In the presence of many magnetic nuclei this minimum will be smeared out.

In another model, RPs were supposed to have many magnetically non-equivalent nuclei. The numerical calculations of the MIE parameter $^T\alpha$ (see Eq. (3.12)) for the H $\rightarrow$ D isotope substitution were made using the continuous diffusion model and the semiclassical description of the hfi. The results are presented in Figs. 4.11 and 4.12.

The two curves in Fig. 4.11 demonstrate that the MIE parameter is smaller for only partially deuterated RPs (Fig. 4.11b) as compared to the complete deuteration of the pair radicals (Fig. 4.11a).

The maximum near 1 mT (see Fig. 4.12) is due to the fact that at these values of B_0, it is mostly the isotropic hfi in the protonated radical A that contributes to S-T conversion. As we see from Fig. 4.12, no additional extreme appears in the field dependence of the magnetic isotope enrichment parameter for RPs with many magnetically non-equivalent nuclei. For such systems, the condition (4.10) only occurs for a

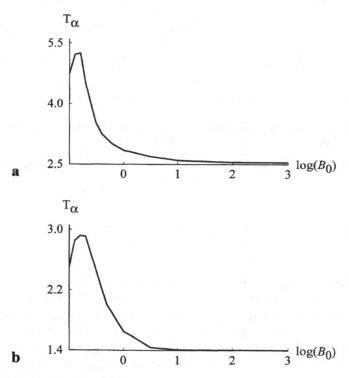

Fig. 4.11. Dependence of the MIE parameter $^T\alpha$ (see Eq. (3.12)) on the field B_0 (B_0 in mT units). In protonated RPs, radicals A and B have the effective hfi constants $(A_{\text{ef}})_A = 1.25$ mT and $(A_{\text{ef}})_B = 2.48$ mT. In deuterated radicals, the effective hfi constants are reduced by the factor four. **a** Curve corresponding to the situation where both radicals are H $\rightarrow$ D isotope substituted, **b** curve corresponding to the RPs where only in radicals B protons are substituted for deuterons. The parameters used for these calculations are: $\Delta g = 0$, $K_S\tau_r = 10$, $J = 0$, and $\tau_D = 1$ ns.

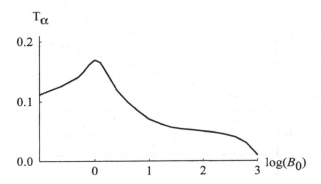

Fig. 4.12. Dependence of the MIE parameter $^T\alpha$ (see Eq. (3.12)) on the field B_0 for H → D isotope substitution (B_0 in mT units). In protonated RPs, radicals A and B have the effective hfi constants $(A_{ef})_A = 1.25$ mT and $(A_{ef})_B = 0.1$ mT. In deuterated radicals, the effective hfi constants are reduced by the factor four. The parameters used for these calculations are: $g_A = 2.0008$, $g_B = 2.0052$, $K_S = 1000$ ns^{-1}, $\tau_r = \tau_D/20$, $\tau_D = 14.8$ ns, and $J = 0$.

negligible fraction of the RPs at any given external field intensity. As expected, MIE vanishes at very high fields, i.e., B_0 about 1 T or more (see Fig. 4.12).

Spin-spin interaction between the two unpaired electrons in RPs may affect the field dependence of the RP recombination probability. This interaction should definitely be taken into account when considering MIE in radical reactions which occur in restricted spaces, like micelles, or in reactions which proceed through biradical intermediates (see also the discussion of this problem in [5, 25, 46]). An interradical spin-spin interaction splits the energy levels of the singlet and triplet states. For instance, Fig. 4.13 schematically shows the field dependence of the RP energy levels in the presence of the Heisenberg exchange interaction.

According to Fig. 4.13, the isotropic hf coupling cannot efficiently mix singlet and triplet states in the low magnetic field region if the exchange integral J (see Eq. (4.9)) exceeds the hf coupling constant. Thus, the exchange interaction can diminish S-T transitions induced by the hfi in the low field region. In the region where S and T_{-1} levels intercept, the hfi is able to induce S-T_{-1} mixing. The level crossing occurs at

$$B_0^* = 2J \ , \tag{4.11}$$

where the exchange integral J is taken in the units of the magnetic field intensity.

Figure 4.14 illustrates the effect of the exchange interaction on the RP recombination. The numerical calculations are done for the two-position model of RP with one magnetic nucleus with spin $I = 1/2$. A triplet precursor of the RP was assumed.

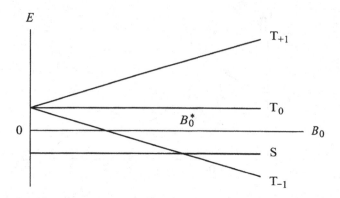

Fig. 4.13. Schematic diagram of the RP spin energy levels in the presence of the Heisenberg exchange interaction. In zero field, the energy levels of singlet and triplet states are split by the exchange interaction. The singlet state level and the triplet T_{-1} state level cross in the field B_0^*.

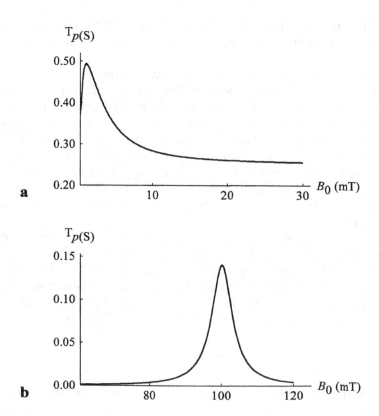

Fig. 4.14. The influence of the RP level crossing on the RP recombination probability. The parameters used for these calculations are: $\lambda = 0.9$, $\tau_2 = 5$ ns, $\tau = 200$ ns, $\Delta g = 0$, $A = 1$ mT, and $I = 1/2$. **a** RP recombination probability without the exchange interaction, $J = 0$. **b** Curve corresponding to the same model of RPs with the exchange interaction, $J = 50$ mT.

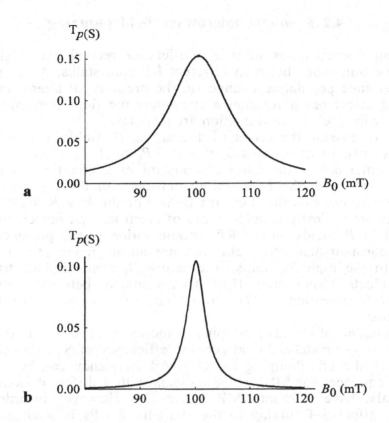

Fig. 4.15. Dependence of the RP recombination probability on the field B_0 for two values of the hf coupling constant A: $A = 2$ mT (**a**) and $A = 0.5$ mT (**b**). Other parameters used for these calculations are: $\lambda = 0.9$, $\tau_2 = 5$ ns, $\tau = 200$ ns, $\Delta g = 0$, and $J = 50$ mT.

From Fig. 4.14b, we see that the triplet-born RP recombination probability passes through the maximum in the level crossing region discussed above. Figure 4.15 represents the RP recombination probability in the level crossing region as a function of the hf coupling constant A. The results of the numerical calculations are exhibited for two values of A.

As suggested by our earlier qualitative discussion, Fig. 4.15 shows that the position of the extremum under study does not depend on the value of the hf coupling constant. But the RP recombination probability is sensitive to the hfi (compare Fig. 4.15a and b). Thus, MIE may be pronounced in the region specified by Eq. (4.11) which determines RP level crossing when the Zeeman interaction and the Heisenberg interaction act simultaneously. In fact, a distribution of the exchange integral is expected for an ensemble of RPs in real systems. As a result, the curves in Fig. 4.15 will broaden.

4.2 Resonant microwave field pumping

Alternating external fields are able to affect RP recombination, since they can induce transitions between different RP spin states. As a result, the RP singlet state population changes in the presence of alternating fields. This field effect has a resonance character: the field frequency should coincide with one of the transition frequencies.

When discussing the effect of alternating B_1 fields, it is convenient to consider two limiting cases $B_0 \leq A$ and $B_0 \gg A$ separately. In the first case, the intensity of the external constant magnetic field is less than or comparable with the hf coupling constant. In the second case, the field intensity exceeds the local hfi field. For the low B_0 fields, the RP spin states are basically coupled states of electrons and nuclei. Until now, the effect of B_1 fields on the RP recombination in the presence of low external constant magnetic fields did not attract much attention in literature. In the high B_0 fields, alternating B_1 fields induce transitions between electron spin states (EPR transitions) or between nuclear spin states (NMR transitions) [27]. Both kinds of transitions affect RP recombination.

The nutation of the nuclear spins in the external resonant field modulates the hfi. Ultimately, this affects the efficiency of S-T mixing in RPs induced by the hfi. Pumping on an NMR frequency can have remarkable consequences for MIE, since radicals with a different isotope composition also have different NMR frequencies. However, in order to significantly affect S-T mixing in the short-lived RPs in such a way, we have to apply high B_1 fields. As a consequence, in the case of the high B_1 fields an excitation of the nuclear spins is expected to be rather nonselective to the isotope composition of radicals.

The most promising way to affect RP recombination by alternating magnetic fields is the pumping of microwave fields in the presence of a constant high field B_0. In this situation, the MW field induces transitions between the RP triplet sublevels. Qualitatively, the effect of the MW pumping on the RP recombination probability can be described by Fig. 4.16.

If RPs start from their triplet states, the hfi in the high field case mixes only the singlet state with one of the triplet states, T_0. Thus, in the absence of the MW field, only 1/3 of RPs, born in the T_0 state, may recombine. The MW pumping changes situation, now RPs born in the T_{+1} and T_{-1} states also have the ability to recombine, since the MW field transfers them to the T_0 state and the hfi further converts them to the reactive singlet state. This resonance MW pumping effect on the RP recombination is successfully used for the indirect detection of EPR frequencies of short-lived RPs (see, e.g., [5, 47, 48]).

To illustrate this qualitative consideration, the B_1 field dependence of the RP recombination probability was numerically calculated for a model RP with one magnetic nucleus of spin $I = 1/2$. The two-position

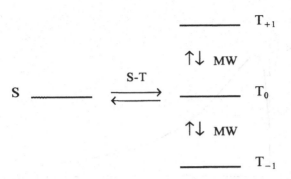

Fig. 4.16. Schematic diagram of the RP spin energy levels in the high magnetic field B_0, of singlet-triplet (S-T_0) transitions. MW field induced transitions between triplet states T_0-T_{+1} and T_0-T_{-1} are shown.

RP kinetic model was used. The spin Hamiltonian of the system in the rotating frame is

$$\mathscr{H} = (g\beta B_0 - \hbar\omega)(S_{Az} + S_{Bz}) + g\beta A\, S_{Az}I_z + g\beta B_1(S_{Ax} + S_{Bx}) \, , \quad (4.12)$$

where ω is the B_1 field frequency, A is the hf coupling constant quoted in the magnetic field units. The ensemble of RPs is divided into two subensembles with the nuclear spin projections $+1/2$ and $-1/2$. For the Hamiltonian (4.12) the EPR frequency of radical B equals $\omega_B = (g\beta/\hbar)B_0$. The radical A has two EPR frequencies: $\omega_A(+) = (g\beta/\hbar)B_0 + (g\beta/\hbar)A/2$ and $\omega_A(-) = (g\beta/\hbar)B_0 - (g\beta/\hbar)A/2$ corresponding to the two projections of the nuclear spin. Using Eqs. (2.87), the RP recombination probability was calculated for both subensembles. During these calculations, it was assumed that the MW field is in resonance with those radicals A which have the EPR frequency $\omega_A(+)$, i.e., $\omega = (g\beta/\hbar)B_0 + (g\beta/\hbar)A/2$. The results are presented in Fig. 4.17.

Two features of the curves in Fig. 4.17 are noteworthy.

For both subensembles of RPs, the recombination probability passes through the maximum with an increase in the MW field intensity. The increase of the RP recombination probability at relatively low values of B_1 confirms our qualitative arguments presented earlier. The decrease of the triplet-born RP recombination probability at the higher values of B_1, when B_1 exceeds the hyperfine splitting of the EPR spectrum, i.e., $B_1 > A$, arises from the spin-locking effect: in the presence of high B_1 fields, both unpaired electron spins precess coherently in the rotating frame and, as a result, RPs conserve their multiplicity, i.e., S-T transitions are suppressed. Further discussion of the spin-locking effect in RP spin dynamics can be found in [5, 48–50].

From the point of view of the magnetic isotope effect, another feature of the curves in Fig. 4.17 is remarkable: the MW field does not

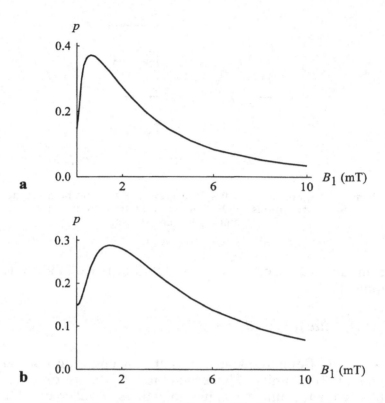

Fig. 4.17. Dependence of the triplet-born RP recombination probability on the field B_1 for two subensembles of pairs with the $+1/2$ (**a**) and $-1/2$ (**b**) projection of the nuclear spin. The field B_1 is in resonance with the EPR transition for the radical with the $+1/2$ projection of the nuclear spin. The parameters used for these calculations are: $A = 2$ mT, $J = 0$, $\lambda = 0.9$, $\tau_2 = 5$ ns, and $\tau = 200$ ns.

equally influence the RP recombination of the two RP subensembles, which differ in their EPR spectra (compare Fig. 4.17a and b). The subensemble of RPs with the $+1/2$ nuclear spin projection is affected by the MW field to a greater extent (Fig. 4.17a) than the subensemble with the $-1/2$ spin projection (Fig. 4.17b). This difference arises from the fact that the MW field frequency was chosen in resonance with the EPR transition for the subensemble with the $+1/2$ nuclear spin projection. Thus, the MW field is able to selectively affect the recombination of the RPs with the particular EPR lines. These arguments can be applied to the MIE. Consider a mixture of RPs with different isotope compositions. Each isotope composition produces a specific hyperfine structure of the EPR spectrum. Therefore, the MW pumping can selectively change the recombination of RPs with a definite isotope composition. By varying the MW frequency or the constant B_0 field intensity, we can either increase or decrease the magnetic isotope enrichment parameter. By the proper MW pumping, the isotopes can be enriched for the specific position of nuclei in molecules.

The possibility of using MW pumping for a diagnosis of the MIE is indeed interesting. The resonant dependence of the isotope effect on MW pumping can thus serve as a reliable criterion of MIE.

From the previous discussion we see that MW pumping is able to affect selectively the magnetic isotope effect. But the interesting possibility is that MW pumping is able to induce MIE even in the absence of MIE without the MW field. In fact, it is not necessary, for a manifestation of the MW field induced MIE, for the hfi to be the main mechanism of singlet-triplet transitions in RPs. Let us suppose that the spin-orbit coupling, but not the hfi, is the dominant mechanism of S-T transitions in RPs. The spin-orbit coupling induces, e.g., the Δg-mechanism of S-T_0 transitions in high magnetic fields or paramagnetic relaxation due to g-tensor anisotropy. In both cases, induced S-T transitions do not depend on the isotope composition of molecules. However, the spin-orbit coupling induced S-T transition rates between the singlet state, on the one hand, and the three triplet sublevels, on the other hand, are different. For instance, the Δg-mechanism induces only S-T_0 transitions, it does not mix the singlet state with the two other triplet states, T_{+1} and T_{-1}. Under this condition, the MW field induced transitions between the triplet sublevels will influence the efficiency of S-T mixing in RPs. The MW field may be in resonance with one of the pair radicals which have a definite isotope composition. And finally, this will lead to MIE since the MW field does not pump radicals with different isotope compositions in the same fashion. Thus, the MW field stimulated magnetic isotope effect can be expected for any reactions proceeding through the radical pair or the biradical intermediate state.

5 Experimental evidences of magnetic isotope effect

This chapter deals with experimental data concerning MIE. It is not our aim to present all experimental results available. The most comprehensive studies of MIE were undertaken for the photolysis of dibenzyl ketone [15, 17, 20, 42, 51–53], as discussed in a review paper [6] and monograph [5] among others. The photolysis of DBK may serve as a model case to demonstrate the basic concepts and predictions of the MIE theory. Therefore, this chapter starts with the consideration of the photolysis of DBK. There are several main trends in the studies of MIE, as outlined throughout this chapter. MIE can be used as a way to determine the mechanism of chemical reactions. Some examples are described below. Quite intriguing is the possibility to use MIE for the separation of isotopes of heavy elements. This and several other potential applications are also presented in this chapter.

5.1 Enrichment of ^{13}C in dibenzyl ketone during photolysis

5.1.1 Photolysis of dibenzyl ketone in homogeneous solutions

The kinetic scheme of the photolytic decomposition of DBK is shown in Fig. 2.1. Decomposition starts from the excited triplet molecule and is followed by radical pair formation: (PhCH$_2$CO $\cdot$ $\cdot$ CH$_2$Ph), with either ^{12}C or ^{13}C in the CO group differing strongly in their effective hf coupling constants. This fact provides favourable conditions for a MIE in the case of ^{12}C $\rightarrow$ ^{13}C isotope substitution in DBK. The hf coupling constants and the S-T dynamics for three different isotope compositions of the RP (PhCH$_2$CO $\cdot$ $\cdot$ CH$_2$Ph) were shown already in Fig. 2.9, demonstrating the fast S-T mixing in RPs with the ^{13}C isotope. Therefore, RPs with ^{13}C are more likely to convert from the initial triplet state to the reactive singlet state and, finally, to recombine, regenerating the starting material, DBK. As a result, the recovered DBK molecules will be enriched in ^{13}C. The magnetic isotope ^{13}C enrichment parameter $^{T}\alpha_1$ (see Eq. (2.12)) for the photolysis of DBK in several homogeneous solutions is given in Table 5.1. In these experiments, two different methods were applied to analyse the isotope effect: mass spectrometry and NMR spectroscopy. Mass spectroscopy provides the more accurate isotope analysis. However, NMR has the advantage of directly monitoring the isotope content at the carbonyl position. Carbonyl ^{13}C splits the resonance of the adjacent methylene protons by 6.3 Hz. Observation of the proton NMR spectrum and comparison of the integrated line area of the ^{13}C satellites to that of the unsplit methylene resonance yields the percentage of ^{13}C [51] and thus the enrichment parameter $^T\alpha_1$.

In non-viscous solvents as hexane or benzene the effective lifetime of the RP in a cage is rather small, $\tau_D \approx 10^{-10}$ s (the estimate is made using Eq. (2.16)). Correspondingly, in these solvents the isotope enrichment parameter is small (see Table 5.1). The results obtained for these solvents cannot be unambiguously qualified as MIE. Indeed, for $^{12}C \rightarrow {}^{13}C$ isotope substitution the familiar mass isotope effect alone is sufficient to yield the enrichment parameter $^T\alpha_1$ as listed in Table 5.1 [1]. However, in more viscous solvents the values of the parameter $|^T\alpha_1|$ presented in Table 5.1 by far exceed the expected values for the mass isotope effect [17]. Thus, the isotope effect due to the photolysis of DBK in viscous solvents (see Table 5.1) must be interpreted as the magnetic isotope effect.

From Table 5.1 we see that the isotope enrichment parameter passes through a maximum with an increase in solvent viscosity. This type of viscosity dependence is consistent with our previous theoretical discussion (see Figs. 3.5, 3.8 and 3.14). The increase of the isotope enrichment parameter with increasing viscosity in the range of $0 < \eta < 10$ P is interpreted as follows. With increasing viscosity the in-cage lifetime of RPs also increases, so that the hfi is able to convert more triplet RPs into the reactive singlet state. The explanation for the decrease of the isotope

Table 5.1. Experimentally measured values of the ^{13}C enrichment parameter $^T\alpha_1$ for the photolysis of DBK molecules in a number of homogeneous solutions with different viscosity of solvents in the Earth's magnetic field.

Solvent	Viscosity (cP)	$^T\alpha_1$	Reference
Hexane	0.3	−0.040	[15]
Benzene	0.6	−0.026±0.002	[17]
Toluene	0.6 (25°C)	−0.038±0.003	[51]
	0.8 (0°C)	−0.061±0.003	[51]
3-Pentanol	4	−0.056±0.002	[51]
Cyclohexanol	60 (25°C)	−0.093±0.002	[51]
	solid (0°C)	−0.112±0.003	[51]
70% cyclohexanol/	380 (−35°C)	−0.166±0.002	[51]
30% isopropanol	1800 (−50°C)	−0.138±0.006	[51]
(w/w)	3200 (−55°C)	−0.105±0.003	[51]
80% cyclohexanol/	3600 (−45°C)	−0.086±0.007	[51]
20% isopropanol	27000 (−60°C)	−0.088±0.008	[51]
(w/w)			

enrichment parameter at very high viscosity (see Table 5.1) is less straight-forward. Several reasons can be given for this decrease. For instance, in solvents with high viscosity, paramagnetic relaxation, independent of the hfi, may contribute significantly to S-T mixing in RPs. Correspondingly, the contribution of the hfi to the RP spin dynamics will be suppressed and MIE will reduce. Another interpretation for this experimental fact is a contribution from a minor channel of RP recombination. According to our discussion in Chap. 3, a pronounced extreme is expected for the viscosity dependence of the isotope enrichment parameter for RPs with many magnetically non-equivalent nuclei, when the RPs are allowed to recombine from both states: the singlet, as a major channel of recombination, and the triplet, as a minor channel of recombination (see Fig. 3.14). In the case shown in Fig. 3.14, the rate of singlet RP recombination, K_S, is 2000 times larger than the rate of triplet RP recombination, K_T. How can a minor channel of RP recombination diminish the magnetic isotope enrichment at high viscosity of the solvent? When solvent viscosity increases, the duration of collision between two RP partners within the recombination zone also increases, and eventually, at high viscosity, the collision duration can be long enough for recombination from both the triplet state and the singlet state. Thus, due to the existence of the minor recombination channel, the singlet-triplet mixing becomes useless for RP recombination. In this case, the isotope composition of molecules does not play a role in the RP recombination, i.e., MIE vanishes.

The theory presented in Chap. 3 provides a satisfactory description of experimental data for the photolysis of DBK. Table 5.2 presents the calculated values of the MIE parameter $^T\alpha_1$ for several values of solvent viscosity. Here, this parameter is defined as

$$^T\alpha_1 = \frac{p\,(\mathrm{DBK}\text{-}^{12}\mathrm{C}) - p\,(\mathrm{DBK}\text{-}^{13}\mathrm{C})}{1 - p\,(\mathrm{DBK}\text{-}^{12}\mathrm{C})} \; , \tag{5.1}$$

where $p(\mathrm{DBK}\text{-}^{12}\mathrm{C})$ and $p(\mathrm{DBK}\text{-}^{13}\mathrm{C})$ are the recombination probabilities for RPs with $^{12}\mathrm{C}$ and $^{13}\mathrm{C}$ isotopes in the carbonyl group of the DBK molecule. Numerical results are obtained with the RP continuous diffusion model using Eqs. (3.30). The effective in-cage lifetime of RPs, τ_D, and their total lifetime in the reaction zone, τ_r, are estimated from the following equations [5]

$$\tau_D = \frac{b^2}{D} \; , \tag{5.2}$$

$$\tau_r = \frac{a\,b}{D} \; , \tag{5.3}$$

where b is the recombination radius, a is the thickness of the recombination layer, and D is the coefficient of mutual diffusion of the two

Table 5.2. Numerically calculated values of $^T\alpha_1$ for $^{12}C \to {}^{13}C$ isotope substitution in DBK.

Viscosity (cP)	Calculations using Eqs. (3.30) and (5.1)	Numerical calculations according to [39]	Calculations using Eqs. (3.30) and (5.1) for the existence of a minor channel of recombination
0.5	−0.028	−0.026	−0.028
1	−0.043	−0.039	−0.043
2	−0.062	−0.055	−0.061
5	−0.095	−0.084	−0.093
10	−0.128	−0.11	−0.122
20	−0.167	−0.14	−0.154
40	−0.213	−0.18	−0.185
60	−0.243	−0.21	−0.199
100	−0.281	−0.24	−0.212
200	−0.335	−0.28	−0.216
400	−0.386	−0.32	−0.204
600	−0.414	−0.34	−0.191
1000	−0.447	−0.37	−0.170
5000	−0.526	−0.43	−0.100

The parameters are: $b = 0.6$ nm, $a = 0.03$ nm, $T = 300$ K, $K_S = 1000$ ns^{-1}, and $K_T = 0.5$ ns^{-1} (the fourth column). Effective hf coupling constant with the protons in the radical PhCH$_2$ equals 1.23 mT. In the radical PhCH$_2{}^{13}$CO the hf coupling constant for ^{13}C is 12.5 mT. Within the framework of the semiclassical description the effective hf coupling constant for ^{13}C is 6.25 mT (see Eq. (2.54)).

partners in the RP. Using Stokes relation, the diffusion coefficient may be expressed through the viscosity η and the temperature

$$D = \frac{2kT}{3\pi\eta b} . \tag{5.4}$$

With $T = 300$ K, $b = 0.6$ nm, and $a = b/20 = 0.03$ nm, inserting Eq. (5.4) in Eqs. (5.2) and (5.3) we obtain

$$\tau_D = 0.245\eta ,$$

$$\tau_r = 0.012\eta , \tag{5.5}$$

where the viscosity η should be taken in cP in order to get parameters τ_D and τ_r in nanoseconds. Results of such calculations of $^T\alpha_1$ are presented in the second column of Table 5.2. For comparison the third column gives values of $^T\alpha_1$ calculated within another approximation: the hf coupling with the protons is described semiclassically, while the interaction with ^{13}C is considered in the consistent quantum mechanical way, not in the semiclassical approximation. For these calculations, the algorithm as outlined in [39] was employed. For the calculations of the fourth column it is assumed that RP recombination from both singlet and triplet states occurs and gives the same final product.

Comparison of Tables 5.1 and 5.2 shows at first glance that the theory describes the experimental data reasonably well. However, we see that the theoretical calculations predict a larger MIE than is observed in experiment. Furthermore, they do not reproduce the extreme in the viscosity dependence of the MIE parameter if the minor channel of RP recombination via the triplet state is not involved in the process (see the fourth column in Table 5.2). In the low viscosity region, when $K_T \tau_r < 1$, the minor channel of RP recombination does not noticeably influence the MIE parameter (compare the fourth column in Table 5.2 with the second column). In the high viscosity region, when $K_T \tau_r > 1$, RP recombination through the triplet state cannot be considered anymore as a minor channel. Its contribution becomes crucial. It is the contribution of the triplet channel to RP recombination that can explain the decrease of the MIE with an increase of solvent viscosity in the high viscosity region.

Nevertheless, this theoretical simulation of the experimental data for the photolysis of DBK is not completely satisfying. As mentioned above, the theoretical model succeeds to reproduce the MIE parameter, i.e., the change of RP recombination probability with isotope substitution. But these calculations do not predict the correct value of the quantum yield for the photolytic decomposition of DBK. Note that the quantum yield ϕ is connected with the RP recombination probability as $\phi = 1 - p$. For the photolysis of DBK molecules in benzene: $\phi = 0.72 \pm 0.12$ [6, 17, 20]. With the parameters, used for the fourth column in Table 5.2, the RP recombination probability and the quantum yield of the DBK decomposition at $\eta = 0.5$ cP are equal to $p(\text{DBK-}^{12}\text{C}) = 0.022$, $\phi(\text{DBK-}^{12}\text{C}) = 0.978$ and $p(\text{DBK-}^{13}\text{C}) = 0.050$, $\phi(\text{DBK-}^{13}\text{C}) = 0.950$. We see that the experimental value of the quantum yield is substantially less than the calculated one. Several reasons can be given for this discrepancy. One possibility is a contribution of the minor channel of RP recombination (via the triplet state). To gain further insight into this problem, let us calculate the magnetic isotope enrichment parameter and the RP recombination probability for the photolysis of DBK for different rates K_T of RP recombination from the triplet state within the reaction radius. The numerical results obtained using Eqs. (3.30) are presented in Fig. 5.1.

From Fig. 5.1 we see that the quantum yield of DBK decomposition in benzene could be attributed to the existence of the minor channel with the RP recombination rate from the triplet state K_T about 30–50 ns^{-1}. These values of K_T do not contradict the experimental data for MIE presented in Table 5.1. Figure 5.2 confirms this statement.

From Fig. 5.2 we see that for $K_T = 30$ ns^{-1} which is 33 times less than the rate of the recombination from the singlet state $K_S = 1000$ ns^{-1} the RP recombination probability is close to 0.2 for $\eta = 0.6$ cP (ben-

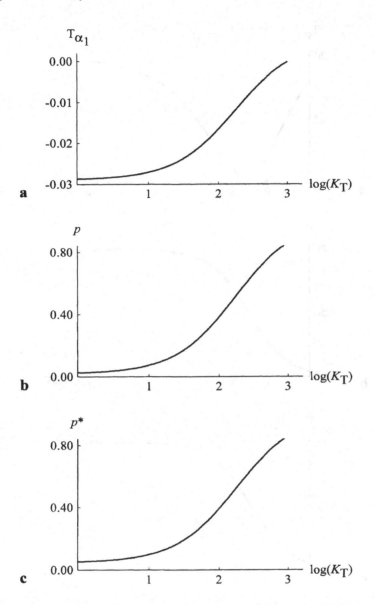

Fig. 5.1. The MIE parameter $^T\alpha_1$ (**a**), recombination probability p of the triplet born pair $(PhCH_2{}^{12}CO\cdot\ \cdot CH_2Ph)$ (**b**, $A_{ef} = 1.23$ mT), and recombination probability p^* of the pair $(PhCH_2{}^{13}CO\cdot\ \cdot CH_2Ph)$ (**c**, $A_{ef} = 8.32$ mT, see Eqs. (3.30)) as a function of the triplet RP recombination rate K_T for the photolysis of DBK (K_T in ns^{-1} units). The parameters used for these calculations are: $\eta = 0.5$ cP and $K_S = 1000$ ns^{-1}.

zene) in accordance with the experimental data, and the viscosity dependence of the MIE parameter $^T\alpha_1$ resembles qualitatively the experimental observation (compare Table 5.1, the third column, with Fig. 5.2a). However, quantitatively we get a new discrepancy: now the magnitude

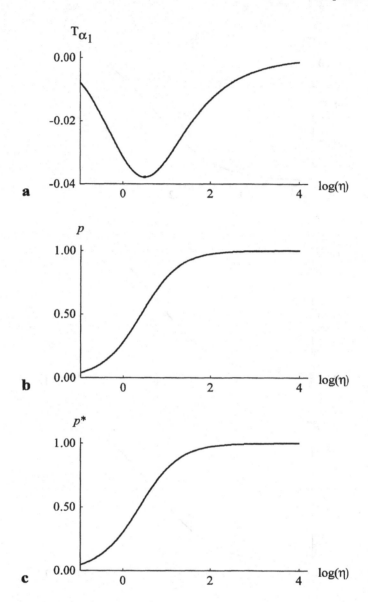

Fig. 5.2. The MIE parameter $^T\alpha_1$ (**a**), recombination probability p of the triplet born pair $(PhCH_2{}^{12}CO\cdot\ \cdot CH_2Ph)$ (**b**, $A_{ef} = 1.23$ mT), and recombination probability p^* of the pair $(PhCH_2{}^{13}CO\cdot\ \cdot CH_2Ph)$ (**c**, $A_{ef} = 8.32$ mT, see Eqs. (3.30)) as a function of viscosity η for the photolysis of DBK (η in cP units). The parameters used for these calculations are: $K_S = 1000$ ns^{-1} and $K_T = 0.03\,K_S = 30$ ns^{-1}.

of the MIE parameter, predicted by the model calculations, is very small even in the case of high solvent viscosity (see Fig. 5.2a). Thus, we see that RP recombination via the triplet channel can describe the experimental data for the quantum yield of the photolysis of DBK in ben-

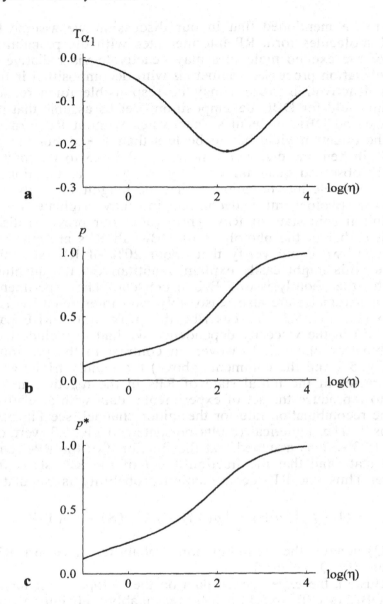

Fig. 5.3. The MIE parameter $^T\alpha_1$ (**a**), recombination probability p of the pair $(PhCH_2{}^{12}CO \cdot \cdot CH_2Ph)$ (**b**, $A_{ef} = 1.23$ mT), and recombination probability p^* of the pair $(PhCH_2{}^{13}CO \cdot \cdot CH_2Ph)$ (**c**, $A_{ef} = 8.32$ mT, see Eqs. (3.30)) as a function of viscosity η for the photolysis of DBK (η in cP units). The parameters used for these calculations are: $f = 0.2$, $K_S = 1000$ ns^{-1}, and $K_T = 0.5$ ns^{-1}.

zene. But then we fail to quantitatively interpret the MIE parameter for solvents with higher viscosity. This interpretation of the photolysis of DBK creates one more inconsistency: it assumes a rather high value of K_T ($K_T = 30$–50 ns^{-1}).

It has to be mentioned that in our discussion we assume that excited DBK molecules form RP intermediates with the probability 1. In fact, part of the excited molecules may deactivate via radiative or nonradiative relaxation processes competing with decomposition into the RP state. This deactivation process may be responsible for a reduction of the quantum yield for DBK decomposition. Let us assume that the fraction x of excited DBK molecules deactivates without RP formation. In this case the quantum yield has to be less than $1 - x$. For the photolysis of DBK in benzene one would need $x = 0.2$–0.3 to interpret the experimentally observed quantum yield by using this assumption.

Another possible interpretation is that the magnitude of the quantum yield for DBK photochemical decomposition may originate from an appropriate initial spin state of RPs. Throughout our previous discussion, we suggested that in the photolysis of DBK all RPs are created in the triplet state. If we could verify that about 20% of RPs start from the singlet state, this might easily explain quantitatively the quantum yield measured for the photolysis of DBK in benzene. The experimental data for the MIE parameter are also reasonably well interpreted by using this assumption (see Fig. 5.3). To describe the peak of the MIE parameter $^{T}\alpha_1$ (Eq. (5.1)) in the viscosity dependence we had to include the minor RP recombination channel. However, in contrast to the previous situation (see Fig. 5.2 and the comments above) the mainly triplet and partly singlet character of the initial state of RPs in the reaction under study allows us to reproduce the set of experimental data with a relatively low value of the recombination rate for the minor channel, see Fig. 5.3 where $K_T = 0.5$ ns^{-1}. The numerical results presented in Fig. 5.3 were obtained using Eqs. (3.30). It is assumed that the fraction f of the RPs starts from the singlet state and that the fraction $1 - f$ of the RPs starts from the triplet state. Thus, the RP recombination probability is calculated as

$$p = (1 - f)\left(^{T}p(S) + ^{T}p(T)\right) + f\left(^{S}p(S) + ^{S}p(T)\right) , \qquad (5.6)$$

where $^{G}p(Q)$ denotes the recombination probability of G-born RPs from their Q state (G, Q = S or T).

In conclusion, the experimental data on the isotope effect for the photolysis of DBK (see Table 5.1) can be reasonably well interpreted in the framework of the RP model paradigm. For this reaction it is well justified to identify the isotope effect as MIE. Further confirmation of MIE in the photolysis of DBK is provided later in this chapter.

5.1.2 Photolysis of dibenzyl ketone in micelles

For reactions inside micelles, the MIE can be much more pronounced as compared to reactions in homogeneous solutions. Photolysis of DBK inside micelles of aqueous detergent solutions was comprehensively studied by N. Turro and B. Kraeutler (see, e.g., [6, 20]). According to [20], ^{13}C-enrichment of DBK recovered from photolysis in aqueous solution of hexadecyltrimethyl ammonium chloride (HDTCl) increases markedly above a "critical micelle concentration" (cmc). Above cmc DBK predominantly exists in the micellar phase. The isotope enrichment parameter equals (see Eq. (2.8))

$$\alpha = \frac{^{12}\phi}{^{13}\phi} = \frac{\text{efficiency of disappearance of } ^{12}\text{C ketone}}{\text{efficiency of disappearance of } ^{13}\text{C ketone}} \, . \tag{5.7}$$

It can be also rewritten as

$$\alpha = \frac{1 - p\,(\text{DBK-}^{12}\text{C})}{1 - p^*\,(\text{DBK-}^{13}\text{C})} \, , \tag{5.8}$$

where p and p^* are the RP recombination probabilities for the two isotope compositions of DBK. In the micellar phase, $\alpha = 1.47$ [20]. A variation of α with the detergent concentration is shown in Fig. 5.4.

A number of conclusions can be drawn from these experimental results. There is a parallelism between the observed value of the isotope enrichment parameter and the micelle concentration. Below cmc

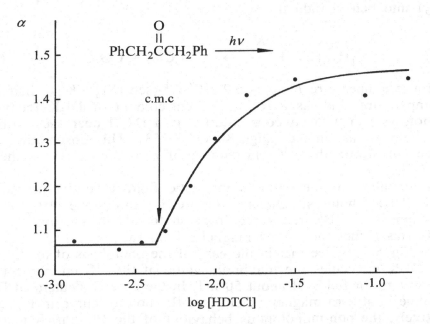

Fig. 5.4. ^{13}C-enrichment of dibenzyl ketone recovered from photolysis in aqueous solution of hexadecyltrimethyl ammonium chloride (HDTCl): plot of the observed enrichment parameter α versus the logarithm of the concentration of the detergent. (Data are taken from [20].)

([HDTCl] $< 8 \cdot 10^{-3}$ M) the reaction occurs in aqueous solution with low viscosity and $\alpha \approx 1.06$. At cmc the parameter α sharply increases. Above cmc, when the reaction proceeds mostly inside micelles, the enrichment parameter α approaches about 1.47 (see Fig. 5.4). Reactions in micelles are characterized by a large cage effect: the quantum yield of photolytic DBK decomposition in micelles is about 0.3 [53]. The fact that at detergent concentrations above cmc, DBK molecules exist and decompose inside micelles was checked by adding the acceptors of radicals, CuCl$_2$ [26, 54]. This acceptor of radicals is soluble only in the aqueous phase. It cannot penetrate micelles, since the cations Cu^{2+} and CuCl$^+$, formed in solution of CuCl$_2$, experience strong repulsion as they approach the positively charged micellar boundary. Thus, if DBK is predominantly dissolved in the micellar phase, then the addition of an acceptor of radicals will not influence the DBK decomposition. This result was observed experimentally: above cmc the quantum yield for the DBK decomposition does not depend on the presence of the acceptor of radicals [26, 54]. Simultaneously, in the range 10^{-3}–10^{-4} M of CuCl$_2$ concentration, the acceptor reduces the production of 1,2-biphenyl ethane (PhCH$_2$CH$_2$Ph). This product is formed in the aqueous phase as a result of the recombination of PhCH$_2$ radicals which escape geminate recombination inside micelles.

For DBK photolysis in micelles, the in-cage lifetime of the primary geminate RP (PhCH$_2$CO $\cdot$ $\cdot$ CH$_2$Ph) is determined by decarbonylation of the ketyl into benzyl radical

$$C_6H_5CH_2CO \xrightarrow{K_{-CO}} C_6H_5CH_2 + CO \; . \qquad (5.9)$$

At room temperature $K_{-CO} = 5.2 \cdot 10^7$ s^{-1} (see [51]). K_{-CO} increases with temperature, and, as a result, ^{13}C-enrichment of DBK recovered from photolysis in 0.05 aqueous solution of HDTCl decreases with increasing temperature in the range 30–60°C [53]. This observation supports the conclusion that ^{13}C-enrichment of recovered DBK is the result of MIE.

With variation of the external magnetic field strength, the isotope effect for DBK photolysis changes. Figure 5.5 shows the efficiency of ^{13}C enrichment of DBK recovered from photolysis in aqueous solution of HDTCl as a function of the magnetic field [55].

From Fig. 5.5 we see that in the case of the photolysis of totally protonated DBK molecules the maximum value of the ^{13}C enrichment parameter occurs for fields of about 50 mT. In the case of deuterated DBK the extreme is at zero magnetic field (see the bottom curve in Fig. 5.5). Qualitatively, the non-monotonous behaviour of the ^{13}C enrichment parameter can be understood with the theoretical arguments discussed in Chap. 4. Indeed, MIE induced by the anisotropic hfi (see Fig. 4.3) and the isotropic hfi (see Fig. 4.11) can reveal an extreme in the field de-

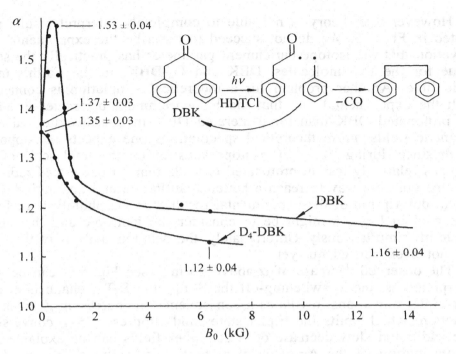

Fig. 5.5. ^{13}C-enrichment parameter α (Eq. (5.7)) as a function of the external magnetic field. (Data are taken from [55].) The top curve corresponds to DBK. The bottom curve corresponds to bis(benzyl-α,α-d$_2$)ketone.

pendence of the parameter α. When the isotropic hfi is the main mechanism of MIE the optimal value of the external field strength should be comparable with the local hfi field (see also Eq. (4.8)). For the photolysis of DBK, this means that the extreme of a function $\alpha(B_0)$ is to be expected at $B_0^* \approx 0.01$ T. The experimental result is quite different. According to Fig. 5.5 (the top curve) the maximum of α occurs at about $B_0^* \approx 0.05$ T. Thus, for the isotropic hfi mechanism of S-T transitions a maximum of the isotope enrichment parameter is expected for the lower values of the external magnetic field compared to the experimental one. Relaxation mechanism of S-T transitions which arise from an anisotropic hfi might be more successfully applied for the interpretation of experimental data under discussion. According to Eqs. (4.2) the relaxation rates decrease for the fields $B_0 > B_0^* \approx 1/\gamma_e \tau_0$, where τ_0 is the correlation time of the radical's rotational motion. Hence, e.g., for $\tau_0 \approx 10^{-10}$ s we have B_0^* about 0.1 T. This estimate is in good accordance with the position of the maximum on the top curve in Fig. 5.5. In Chap. 4 it was also demonstrated that depending on kinetic and magnetic resonance parameters the isotope enrichment parameters can either vary monotonously (see Fig. 4.2) or non-monotonously (see Fig. 4.3). Thus, the theory of MIE might provide an interpretation for the field dependence of the isotope enrichment parameter.

However, the theory is not able to completely interpret results presented in Fig. 5.5. We do not succeed to describe the experimental observation that the isotope enrichment parameter has practically the same value for the two molecules, DBK and D_4-DBK, in the Earth's magnetic field. Another problem for a theoretical simulation is connected with the experimental fact that the ^{13}C enrichment parameter is larger for protonated DBK than for deuterated DBK in intermediate and high magnetic fields. From theoretical speculations one expects the opposite result since during $^{12}C \rightarrow {}^{13}C$ isotope substitution the total scale of hfi changes relatively less in protonated radicals than in deuterated radicals.

One possible way to reach a better quantitative and qualitative theoretical description of the experimental results for the photolysis of DBK presented in Fig. 5.5 might be to consider the isotropic and the anisotropic hfi simultaneously. Unfortunately, the detailed analysis of this type has not been carried out yet.

The observed decrease of α above 50 mT (see Fig. 5.5) can be well interpreted as due to switching-off the $S\text{-}T_{+1}$ and $S\text{-}T_{-1}$ channels of singlet-triplet transitions in RPs as a consequence of increasing Zeeman interaction, which splits the triplet states and suppresses S-T conversion. The additional slow decrease of α at higher fields can be explained by a contribution of the Δg-mechanism to $S\text{-}T_0$ transitions. In very high magnetic fields, the Δg-mechanism contribution dominates any variation of $S\text{-}T_0$ transitions induced by hyperfine interaction.

Next we compare data for DBK photolysis in homogeneous solutions (Table 5.1) with those in micellar solutions of DBK (see Table 5.3).

The magnitude of the ^{13}C enrichment for reactions in micelles is significantly larger than for reactions in homogeneous solutions with similar viscosity. Indeed, the microviscosity of the micellar phase is comparable to the viscosity of cyclohexanol. But in the Earth's magnetic field, the reaction in cyclohexanol gives $\alpha_1 = -0.093$ while $\alpha_1 = -0.32$ for HDTCl. This observation shows that the increase in magnitude of the MIE in micellar solutions when compared to the MIE in homogeneous solutions is not related to an increase of the micellar microviscosity but rather to a larger lifetime of the RPs in the micellar super cage.

Table 5.3. Experimental magnitudes of the ^{13}C enrichment parameter $^T\alpha_1$ (Eq. (5.1)) for DBK photolysis in micelles for hexadecyltrimethyl ammonium chloride with microviscosity inside the micellar phase in a range 30–100 cP [17].

B_0 (T)	$^T\alpha_1$
0	−0.32
1.45	−0.13
10	−0.02

MIE in micellar solution was also comprehensively studied for the photolysis of different derivatives of DBK and other ketones. The results are nicely described in a review paper [6] and can be summarized with the following statements. The ^{13}C enrichment of DBK substituted at sites other than carbonyl group demonstrates good correlation with the ^{13}C hyperfine coupling constant at the respective site [56]. MIE for H → D isotope substitution is much less than for ^{12}C → ^{13}C isotope substitution [26]. The studies of the photolysis of DBK were also extended to polymer films (e.g., polymethylmetacrylate), $\alpha = 1.30$, and porous glass, $\alpha = 1.22$ [57]. These ^{13}C enrichment parameters are comparable with those in micellar solution. The large magnitudes of MIE in these systems are the consequence of "restricted space" for diffusion realized by micellar cavities, fluid portions of the polymer films, or cavities of porous glass.

5.1.3 Magnetic isotope effect in photoinduced emulsion polymerization

Oil-soluble ketones (for instance, dibenzyl ketone or its derivatives) can be used as initiators of photochemically induced emulsion polymerization [26, 58]. In [58], emulsion polymerization of styrene, producing polystyrene, was found to depend on the ^{13}C content in DBK. DBK is the photoinitiator of this process. In this case, monomers and initiator are dissolved in the micellar phase. Photodecomposition of DBK produces radicals which initiate the polymerization. Recombination of these radicals inhibits the development of chain polymerization. During the photolysis of DBK, RPs are created in the triplet state. The conversion of RPs from the triplet state to the singlet increases the RP recombination probability. Thus, it is expected that increasing the S-T transition rate will decrease the length and average molecular weight of polymer units. The efficiencies of three photoinitiators, differing in their isotope composition, were compared [58]: DBK, C$_6$H$_5$^{13}CH$_2$CO^{13}CH$_2$C$_6$H$_5$ (DBK-2,2'-^{13}C, 90%) and C$_6$H$_5$CH$_2$^{13}COCH$_2$C$_6$H$_5$ (DBK-1-^{13}C, 90%). The efficiency of (DBK-1-^{13}C, 90%) as a photoinitiator is the same as that of DBK, while efficiency of (DBK-2,2'-^{13}C, 90%) as a photoinitiator is less than that of DBK. The molecular weight distribution of polymer depends on the ratio of DBK and (DBK-2,2'-^{13}C, 90%): in general, the polymer molecular weight distribution exhibits two peaks (see Fig. 5.6).

This figure shows that the relative peak intensity is strongly influenced by the DBK isotope composition. This can be interpreted in the following way. After the decomposition of DBK, the decarbonylation of PhCH$_2$CO radical occurs rather quickly. As a result, a secondary pair of two equivalent phenyl radicals is formed. These phenyl radicals are known to be the initiators of polymerization. Therefore, the primary pair (C$_6$H$_5$CH$_2$CO· ·CH$_2$C$_6$H$_5$) does not influence the MIE. The MIE arises

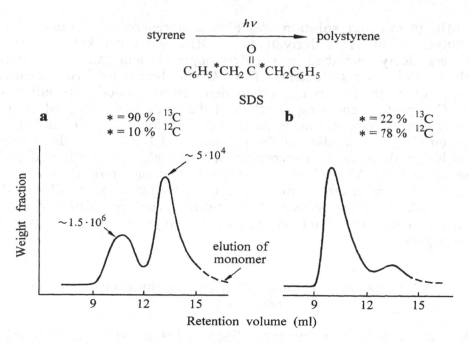

Fig. 5.6. Size-exclusion chromatograms of polystyrene produced by emulsion polymerization of styrene with different mixtures of DBK and DBK-2,2'-^{13}C as photoinitiators. The molecular weight distribution of the polymer produced by photoinitiation with the two mixtures of DBK and DBK-2,2'-^{13}C shows a clear ^{13}C isotope effect. (Data are taken from [58].)

only from the secondary pair ($C_6H_5CH_2 \cdot \cdot CH_2C_6H_5$) of phenyl radicals. Thus, we can elucidate the different outcomes for the two DBK mixtures.

In experiments discussed the content of DBK-2,2'-^{13}C and DBK-1-^{13}C molecules was relatively small. As a consequence, a very small fraction of micelles can possess two or more DBK molecules and the average number of DBK molecules per micelle is less than one.

The bimodal molecular weight distribution is attributed to the existence of two kinds of micelles: in one of them only one DBK molecule is dissolved, while in the other only one DBK-2,2'-^{13}C molecule is dissolved. Hyperfine interaction with ^{13}C provides a more efficient conversion of RPs to the reactive singlet state and as a result in the micelles containing DBK-2,2'-^{13}C molecule polymerization is expected to be less efficient. Thus, in the micelles with DBK molecule, the polymer product will have a higher molecular weight than in the micelles with DBK-2,2'-^{13}C molecule. Therefore, the bimodal molecular weight distribution appears. This bimodality disappears in the presence of a high external magnetic field because in high fields the Δg-mechanism, which is independent of the isotope composition, will dominate the singlet-triplet evolution of RPs [58] and hence will suppress the contribution of the hyperfine interaction to S-T$_0$ transitions.

5.2 Magnetic isotope effect in radiolysis of aromatic hydrocarbon solutions in alkanes

Spin correlation effects in radiolysis of aromatic hydrocarbons in alkanes are well proven and have been extensively studied by B. Brocklehurst [9, 59–61]. Radiolysis of aromatic hydrocarbon solutions produces spin-correlated ion pairs. Geminate recombination of the solute ions M^+ and M^- is the major primary process of radiolysis of these solutions and generates an excited solute molecule. The kinetic scheme of the primary processes of radiolysis for an aromatic hydrocarbon solute, M, in an alkane solvent, S, is shown in Eqs. (5.10)–(5.15)

$$S^* \longrightarrow S^+ + e^- , \tag{5.10}$$
$$S^+ + M \longrightarrow M^+ + S , \tag{5.11}$$
$$e^- + M \longrightarrow M^- , \tag{5.12}$$
$$M^+ + e^- \longrightarrow M^* , \tag{5.13}$$
$$S^+ + M^- \longrightarrow M^* + S , \tag{5.14}$$
$$M^+ + M^- \longrightarrow M^* . \tag{5.15}$$

At high concentrations of M, reaction (5.15) appears to be a major source of the excited solute molecules. Pairs of ions $(M^+ ... M^-)$ are formed in the spin correlated state. They inherit their multiplicity from excited solvent precursor molecules. The fluorescence of M^* can be used to monitor the magnetic field and magnetic isotope effects in the reaction (5.15). This specific reaction is expected to be suitable to reveal MIE, since a pair of ions $(M^+ ... M^-)$ has a long in-cage lifetime. Indeed, in nonpolar solvents the Onsager radius is about 30 nm at room temperature. Therefore, in nonpolar solvents, ion pairs $(M^+ ... M^-)$ have 10–100 ns for their geminate recombination [9, 59–61]. During this relatively long time even a weak hyperfine interaction with coupling constants about 0.1–1 mT can succeed to efficiently mix singlet and triplet states of ion pairs. As a result, the recombination luminescence is predicted to be sensitive to the isotope composition of the solute molecules.

Variation of fluorescence, arising from the H → D isotope substitution during radiolysis of aromatic molecules in non polar solvents is experimentally established [60, 62]. For instance, in [60] pulse radiolysis of solutions of para-terphenyl in decalin, squalane and benzene has been studied. Time-resolved measurements of the recombination luminescence were done using single-photon counting. In this case spin-correlated RPs $(M^+ ... M^-)$ are created in the singlet state. Spin dynamics induced by hfi mixes S and T states. As a matter of fact, reaction (5.15) can proceed for both singlet and triplet states of ion pairs. However, recombination from S and T states can easily be distinguished. Recombination from the triplet state gives excited triplet molecules. They deactivate predominantly via non-radiative relaxation processes. Recombination of singlet pairs gives excited singlet molecules. They are responsible for the observed fluorescence. The ratio R of the fluorescence intensity for two values of the external field was measured, in the presence of a high magnetic field 0.16 T and with zero external magnetic field. In the case of protonated terphenyl molecules the ratio R is larger than in the case of deuterated terphenyl molecules. This observation is in accordance with the decrease of the hfi for the H → D isotope substitution (see Chap. 2, e.g., Eq. (2.55) or Fig. 2.9)). Thus, time-resolved measurements of the recombination luminescence during radiolysis of terphenyl in decalin confirm the concepts of MIE. Furthermore MIE provides an alternative approach to study the primary processes of radiolysis (see also the discussion of this problem in [5, 63]).

5.3 Magnetic isotope effect for heavy elements

So far, MIE was studied mostly for H → D and ^{12}C → ^{13}C isotope substitution. Extension to other elements, especially heavy atoms, seems especially interesting. In the case of heavy atoms, the role of the spin-orbit coupling, as a driving force of the RP singlet-triplet conversion, becomes more important. As a result, it must be checked whether hyperfine interaction is able to compete with spin-orbit coupling (SOC) in S-T mixing. As a matter of fact, SOC effect has already been discussed previously in the context of non-zero Δg, which arises from SOC [27, 64]. It was demonstrated several times that MIE is reduced by the Δg-mechanism for singlet-triplet transitions or by S-T transitions induced by the paramagnetic relaxation due to g-tensor anisotropy.

Nevertheless, the effect of hyperfine interaction on the reaction yield has been observed for various heavy elements: nitrogen [65, 66], oxygen [57, 67–69], silicon [70, 71], sulphur [72, 73], germanium [71, 74–76], and uranium [18, 77, 78].

Nitrogen. The magnetic field dependence of the delayed fluorescence and of the injection current was investigated [66] for an anthracene crystal with dye molecules tetraethylrhodamine (rhodamine B) absorbed on the surface. Photoinduced electron transfer from anthracene (A) to the excited dye (D) creates singlet radical ion pairs, $^1(D^-...A^+)$, on the surface of the crystal. The intensity of the delayed fluorescence and the injection current are determined by the singlet-triplet transitions in these pairs. The experiments have shown that the measured quantities vary with the magnetic field strength in a way which is expected for the magnetic field dependence of S-T transitions induced by the hfi. The field dependence of the delayed fluorescence (or of the injection current) is characterized by the field parameter $B_{1/2}$ defined according to the equation

$$\phi_T(B_{1/2}) = \frac{1}{2}(\phi_T(0) - \phi_T(\infty)) \,, \qquad (5.16)$$

where $\phi_T(B)$ is the triplet yield for the respective field B. Analogously $B_{1/2}$ is determined for the field dependence of the injection current. Both observables turn out to yield the same magnitude of $B_{1/2}$ value. It was experimentally shown that $B_{1/2}$ correlates with the hyperfine coupling constants of nitrogen. Two rhodamine B molecules, differing in their isotope composition, were employed: $C_{22}H_{22}{}^{14}N_2O_3$ and $C_{22}H_{22}{}^{15}N_2O_3$. With $^{14}N \to {}^{15}N$ isotope substitution $B_{1/2}$ decreases from $B_{1/2}(^{14}N) = 2.3$ mT to $B_{1/2}(^{15}N) = 1.86$ mT in agreement with the corresponding decrease of the effective hf coupling constant. Deuteration of the anthracene molecules did not change the field dependences of the delayed fluorescence and the injection current. At a first glance, this experimental observation may be surprising, since deuteration reduces the scale of the hfi four times. However, averaging of the hf coupling occurs due to the fast hole hopping among anthracene molecules and explains the missing magnetic isotope effect for $H \to D$ isotope substitution.

Oxygen. An example for magnetic isotope enrichment in an oxidation chain reaction is given in [67]. Another example of MIE involving oxygen will be presented later, when we discuss MIE for reactions which proceed via a biradical state. The oxidation chain reaction of polymers is terminated by the recombination of peroxy radicals RO_2. The product of the recombination is an unstable tetroxide intermediate. It decomposes and yields oxygen molecules. The kinetic scheme is as follows:

$$2RO_2 \longrightarrow \{RO_2 \cdot \ \cdot RO_2\} \longrightarrow RO_4R \longrightarrow O_2 + \text{products} \,. \quad (5.17)$$

In this case, the diffusion RPs form the singlet and triplet state with the ratio $1 : 3$. The hfi with ^{17}O accelerates the singlet-triplet transitions in those RPs which contain an ^{17}O terminal atom in one (or both) radicals. Therefore, the radicals with ^{17}O recombine with a higher probability than the radicals with ^{16}O and ^{18}O. As a consequence, oxygen product molecules will be enriched in ^{17}O.

Silicon. Magnetic isotope selection for ^{29}Si has been observed in the photolysis of a silyl-containing ketone [70]. Ketone $PhCH_2COSi(CH_3)_2Ph$ photodecomposes through the singlet state in the case of direct photolysis:

$$PhCH_2COSi(CH_3)_2Ph \underset{}{\overset{h\nu}{\rightleftarrows}} {}^1\{PhCH_2\cdot \ \cdot COSi(CH_3)_2Ph\}$$

$$\downarrow$$

$$\text{(products)} \,. \qquad (5.18)$$

With triplet sensitizers, like acetophenone and triphenylene, ketone decomposition starts with the creation of triplet RPs:

$$hv$$
$$PhCH_2COSi(CH_3)_2Ph \rightleftharpoons {}^3\{PhCH_2\cdot \;\cdot COSi(CH_3)_2Ph\}$$
$$\downarrow$$

$$(products)\;. \hspace{4cm} (5.19)$$

For direct photolysis, the silicon isotope effect was indistinguishable from the mass isotope effect. For triplet sensitized photolysis, the isotope enrichment parameter ${}^T\alpha({}^{29}Si)$ equals 1.07, which is definitely larger than the mass isotope effect. Significantly smaller isotope enrichment consistent with the mass isotope effect was detected for the ${}^{30}Si$ nonmagnetic isotope. The ${}^{29}Si$ enrichment in the recovered $PhCH_2COSi(CH_3)_2Ph$ during the triplet sensitized photolysis is attributed to the magnetic isotope effect.

Sulphur. A triad of isotopes: ${}^{32}S$ $(I = 0)$, ${}^{33}S$ $(I = 3/2)$ and ${}^{34}S$ $(I = 0)$ is available. This circumstance provides a good opportunity to discriminate between mass and magnetic isotope effect. MIE for sulphur has been detected during photolysis of phenacylsulphone $(PhCOCH_2SO_2Ph)$ in aqueous solution of sodium dodecylsulphate (SDS) [72, 73]. According to [72], photolysis proceeds from the lowest triplet state of phenacylsulphone by β scission:

$$hv$$
$$PhCOCH_2SO_2Ph \rightleftharpoons {}^3\{PhCOCH_2\cdot \;\cdot SO_2Ph\}$$
$$\downarrow$$

$$PhCOCH_3 + PhSO_2H\;. \hspace{3cm} (5.20)$$

Primary triplet RPs can recombine to the extent that they will be converted to the singlet state. Studies of the magnetic field dependence of the regeneration of the starting molecules via recombination of primary RPs and analysis of the chemically induced nuclear and electron spin polarization confirm [72, 73] that the isotropic hyperfine coupling noticeably contributes to the singlet-triplet mixing in the RP $(PhCOCH_2\cdot \;\cdot SO_2Ph)$. Therefore, the geminate recombination of this RP is a good candidate to observe MIE for sulphur. In fact, it was found [73] that the recovered (starting) molecules are enriched with the magnetic isotope ${}^{33}S$. The enrichment parameter $\alpha({}^{33}S) = 1.015$. We see that this enrichment parameter is considerably smaller than that for ${}^{13}C$ in the micellar photolysis of dibenzyl ketone. This fact may be somewhat surprising. The isotropic hf coupling constant of ${}^{33}S$ in phenylsulphonyl radical is 8.3 mT, hence a large MIE is expected for this reaction. However, in the case of sulphur atoms, spin-orbit coupling may contribute more to the singlet-triplet mixing than in the case of carbon atoms. Being independent of the isotope composition, the spin-orbit coupling contribution to the RP spin dynamics will reduce the effect of the hyperfine interaction (see also the discussion of this problem in [73]). Despite the fact that ^{33}S enrichment is relatively small, it can be definitely attributed to MIE. If the ^{33}S enrichment would be due to the mass isotope effect, the isotope enrichment should be more effective for ^{34}S. But the experimental results show the opposite: ^{34}S enrichment is considerably less than that for ^{33}S. This observation supports the magnetic nature of sulphur isotope enrichment during photolysis of phenacylphenylsulphone in micellar solutions [73].

Germanium. For such a heavy atom as Ge, spin-orbit coupling is expected to dominate the RP singlet-triplet mixing and the hyperfine interaction contribution should be largely suppressed. Thus, it is a priori not clear whether magnetic isotope selection may be observed for Ge or not. But there are interesting experimental indications that germanium can be considered as a potential candidate for magnetic isotope selection. For instance, in [71, 74], the magnetic field dependence of the recombination of RPs involving germyl radicals was studied. The magnetic field dependence of the RP recombination shows a behaviour as expected for singlet-triplet transitions induced by hyperfine interaction. Despite the pronounced spin-orbit coupling, the hyperfine coupling with the magnetic isotope ^{73}Ge may reveal itself in the RP spin dynamics. To succeed with a magnetic isotope selection or magnetic isotope enrichment for such heavy atoms as Ge, it is important to find a suitable reaction system (see the discussion of this problem in [71]).

The successful study on the enrichment of ^{73}Ge with the MIE of Ge-centered radicals was undertaken recently in [75, 76]. MIE was studied by the photolysis of methyltriphenylgermane (Ph_3MeGe) in a polyoxyethylene dodecyl ether micellar solution. Isotope ratios of $^{72}Ge/^{74}Ge$ and $^{73}Ge/^{74}Ge$ were measured using plasma mass spectrometer before and after photolysis. Under a light irradiation excited triplet state of Ph_3MeGe is produced, and then $(^3Ph_3MeGe)^*$ decomposes to a triplet radical pair of the diphenylmethylgermyl ($Ph_2MeGe\cdot$) and phenyl ($Ph\cdot$) radicals in micellar supercage. The triplet-singlet conversion of the radical pair involving magnetic ^{73}Ge ($I = 9/2$) is much faster than that involving nonmagnetic Ge. From the singlet radical pair the cage recombination to the starting compound can occur. Thus, the ^{73}Ge isotope can be enriched in the starting compound. Finally, a significant enrichment of magnetic ^{73}Ge in contrast to little change of nonmagnetic ^{72}Ge is expected. The isotope enrichment parameter observed was $^T\alpha_1 = -0.006 \pm 0.0007$. This value is much smaller than those obtained for C-centered radicals. This decrease in the MIE of ^{73}Ge can be attributed to the spin-orbit interaction of Ge which enhances the nuclear spin-independent triplet-singlet conversion of the radical pair. At the same time, the MIE parameter observed for the photolysis of Ph_3MeGe considerably exceeds that expected from the mass isotope effect ($^T\alpha_1 = -0.0015$). The magnetic field dependence of the isotope enrichment was also measured in [75]. For ^{73}Ge there is a maximum of the isotope enrichment at 0.02 T. This magnetic field dependence of the isotope enrichment of ^{73}Ge is strong evidence for the enrichment of ^{73}Ge due to the MIE.

Uranium. The separation of uranium isotopes, ^{235}U (isotope content of natural uranium 0.72%, nuclear spin $I = 7/2$, magnetic moment $\mu = -0.31\mu_N$), and ^{238}U (nonmagnetic isotope) is an intriguing problem. For this pair of isotopes, the square root of the ratio of their masses is 1.0064. Therefore, the mass isotope effect is very small, and it is highly desirable to find a way to exploit the magnetic isotope separation for uranium. There are several promising attempts – with positive results – to solve this problem [18, 77, 78]. For instance, for uranyl nitrate photoreduction by p-methoxyphenol (ArOH), magnetic isotope enrichment with $\alpha = 1.02$ was obtained [18]. The following reaction scheme is generally accepted [18]:

$$UO_2^{2+} + ArOH \longrightarrow {}^3[UO_2^+ \cdots ArO] + H^+ , \qquad (5.21)$$

$$^3[UO_2^+ \cdots ArO] \longrightarrow UO_2^+ \cdot + \cdot ArO , \qquad (5.22)$$

$$^3[UO_2^+ \cdots ArO] \longrightarrow UO_2^{2+} + ArOH , \qquad (5.23)$$

$$UO_2^+ + UO_2^+ + 4HF \longrightarrow UO_2^{2+} + UF_4 \downarrow + 2H_2O . \qquad (5.24)$$

According to this scheme, the reaction (5.23) regenerates the starting uranyl. This reaction requires $T \rightarrow S$ conversion of the RPs. The hyperfine interaction with ^{235}U may assist in this conversion. When the hf coupling with ^{235}U plays any noticeable role in the spin dynamics in the RP, then the recovered uranyl will be enriched in the isotope ^{235}U, while the precipitated UF_4 should be depleted by ^{235}U. The experiment gave the following results. In the starting uranyl nitrate, the $^{238}U/^{235}U$ ratio, R, was $R = 8.93 \pm 0.01$, while in the reaction product UF_4, R was in the range of $(8.970 \pm 0.005) - (8.945 \pm 0.008)$ [18, 78]. The isotope effect observed is rather small, but it definitely exceeds the mass isotope effect. The sign of the isotope effect observed agrees with the sign expected for the magnetic isotope separation in this case.

5.4 Magnetic isotope effect for biradical reaction pathways

Biradicals attract much attention from the point of view of MIE. Singlet-triplet transitions in biradicals are one of the reaction rate-determining processes for the decay of triplet biradicals [79–83]. Depending on the configuration of the biradical, the singlet-triplet conversion can correspond either to nonradiative intersystem crossing within molecules or to the singlet-triplet dynamics in RPs as described above. Different configurations of biradicals are specified by different magnitudes of the exchange integral. For biradicals with a large exchange integral, the contribution of the isotope selective hyperfine interaction to singlet-triplet transitions is expected to be suppressed. In contrast, the exchange integral for extended configurations, e.g., of flexible chain connected biradicals is relatively small, and biradicals with these configurations can manifest spin dynamics similar to that in RPs [79–83], i.e., hfi can reveal itself in the biradical spin dynamics. This statement is supported by the observation of chemically induced nuclear spin polarization in the course of biradical reactions [79–81]. Therefore, magnetic isotope effect can be expected for biradical reactions. Two examples will be described here: the ^{13}C enrichment in products of the photolysis of 2,n-diphenylcycloalkanones (n-membered ring, $n = 10, 11, 12, 15$) [84] and the isotope selection for oxygen in thermolysis of endoperoxides [6, 57, 69].

Substantial ^{13}C enrichment was found for photolysis of 2,n-diphenylcycloalkanones in hexane [84]. The reaction scheme is presented in Fig. 5.7.

It is assumed that for small distances between unpaired electrons, i.e., strong exchange interaction, the singlet-triplet transitions are induced by spin-orbit coupling, while for configurations with large distances between unpaired electrons the hyperfine interaction can contribute significantly to the singlet-triplet dynamics. As a result of the S-T conversion induced by hf coupling, all carbonyl containing molecules (recovered starting

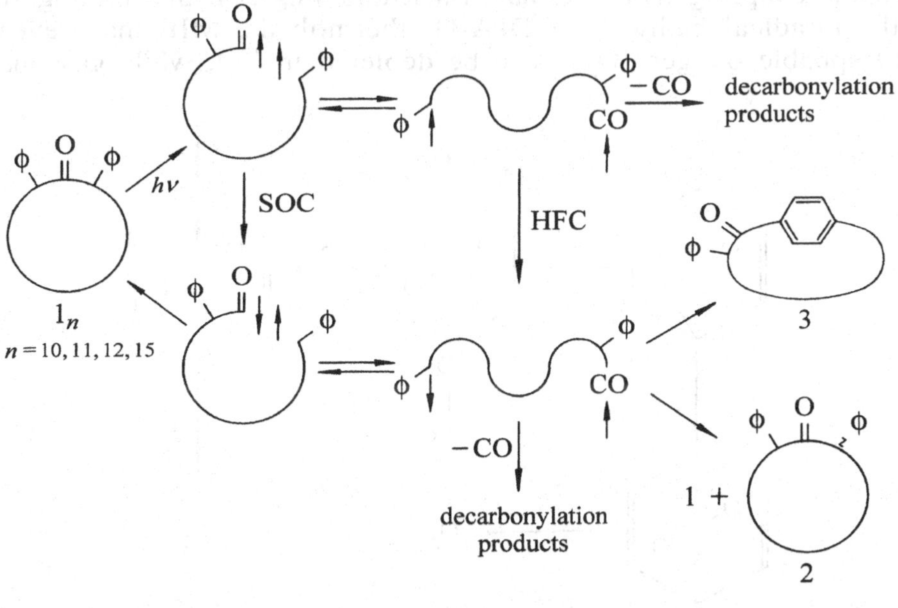

Fig. 5.7. Kinetic processes in biradicals produced by photolysis of 2,n-diphenylcyclo-alkanones. (Scheme is taken from [84].)

molecules, products 2 and 3 in Fig. 5.7) are enriched in ^{13}C, and CO is enriched in [12]C. These experimental results are interpreted as MIE. Indeed, triplet biradicals need to convert from the triplet to singlet state to be able to recombine hence recovering the starting material, or to recombine and thus giving products 2 and 3. Hyperfine coupling with ^{13}C increases the efficiency of S-T mixing in those biradicals possessing [13]C. For this system, the ^{13}C enrichment observed is large and comparable with that for the photolysis of DBK in micellar solutions. It was found that the ^{13}C enrichment depends on temperature and ring size. With this it becomes possible to determine the dynamical pathways through which each product is formed [84].

MIE can be very helpful in discriminating the biradical pathway of a reaction. As an example, let us consider the results obtained for the thermolysis of endoperoxide, 9,10-diphenylanthracene (DPA-O_2) dissolved in CHCl$_3$ and dioxan [6, 57, 69]. This reaction produces oxygen in the singlet state ($^1O_2$) and in the triplet state ($^3O_2$). The reaction scheme is presented in Fig. 5.8.

Singlet oxygen is highly reactive. Thus, by adding quenchers of $^1O_2$, the yield of gaseous $^3O_2$ and the amount of the trapped molecules $^1O_2$ could be determined [6, 57, 69]. It was found that thermolysis of endoperoxides produces oxygen via two pathways. One way is the biradical mechanism (reactions a, b, c and d in Fig. 5.8). Another way is a "concerted mechanism" (reaction e in Fig. 5.8), which produces only singlet oxygen molecules. The biradical mechanism gives both singlet and triplet oxygen molecules due to singlet-triplet transitions in the biradical state. Hyperfine coupling with [17]O may assist this singlet-triplet mixing. Thus, for the biradical pathway of DPA-O_2 thermolysis, MIE may reveal itself: trappable oxygen $^1O_2$ has to be depleted in [17]O, while gaseous $^3O_2$

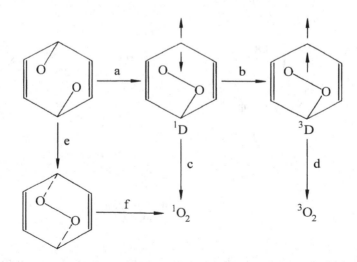

Fig. 5.8. Thermolysis of endoperoxides.

has to be enriched in ^{17}O. Experimental data confirm these theoretical predictions. Moreover, DPA-O$_2$ thermolysis depends on the strength of the external magnetic field. In magnetic fields about 1 T, isotope separation for DPA-O$_2$ decreases. This is due to the contribution of the Δg-mechanism to the singlet-triplet transitions. For the biradicals under consideration, $\Delta g \approx 0.01$. In high magnetic fields, the Δg-mechanism of S-T conversion suppresses the contribution of the isotope selective hf coupling mechanism. The consequence is a decrease of the isotope selection in high fields. In the experiments described, a negligible mass isotope effect was detected: no noticeable selection or enrichment in ^{16}O and ^{18}O was found.

5.5 Magnetic isotope effect in reaction rates

Singlet-triplet transitions induced by hfi will affect RP recombination kinetics. A striking effect is the oscillation of the recombination luminescence [60, 85, 86]. As pointed out in Sect. 5.2, Brocklehurst [60] observed different time dependences of the fluorescence intensity during pulse radiolysis of 0.005 M terphenyl in decaline for H $\rightarrow$ D isotope substitution: the luminescence intensity rises more quickly for terphenyl-h$_{14}$ than for terphenyl-d$_{14}$, in correspondence to the hf coupling constants. Singlet-triplet oscillations are more pronounced in the case of terphenyl-h$_{14}$. Laser-photolysis studies of the MIE on the dynamical behaviour of transient intermediates were performed by Sakaguchi et al. [87]. They investigated the lifetime of benzophenone ketyl radicals formed during benzophenone (BP) photolysis in sodium dodecyl sulfate (SDS). BP is dissolved in the micelles, i.e., it is surrounded by detergent molecules RH. In the concentration range of their experiments, the average number of BP molecules per micelle was less than one. BP photolysis proceeds through the following sequence of processes:

$$BP \longrightarrow {}^{1}BP^* , \tag{5.25}$$

$$ {}^{1}BP^* \longrightarrow {}^{3}BP^* , \tag{5.26}$$

$$ {}^{3}BP^* + RH \longrightarrow {}^{3}(K \cdot \cdot R) , \tag{5.27}$$

$$ {}^{3}(K \cdot \cdot R) \rightleftarrows {}^{1}(K \cdot \cdot R) , \tag{5.28}$$

$$ {}^{3,1}(K \cdot \cdot R) \longrightarrow K \cdot + \cdot R , \tag{5.29}$$

$$ {}^{1}(K \cdot \cdot R) \longrightarrow K\text{--}R, BP + RH . \tag{5.30}$$

The transient absorption spectra show a peak at 525 nm. This absorption is due to $^{3}BP^*$, $^{3,1}(K \cdot \cdot R)$, and K. The intensity of this absorption decreases in two steps: there are fast and slow components in the decay curves. The fast component disappears within a few microseconds. The rate, K_f, of this fast process is sensitive to the isotope composition of BP and to the presence of the external magnetic field. This fast de-

Table 5.4. MIE and magnetic field dependence of the absorption decay rate K_f in micellar SDS solutions of BP. This rate characterizes the decay of absorption intensity at 525 nm [87].

	B_0 (mT)	BP	BP-d_{10}	BP-^{13}C
K_f (10^6 s^{-1})	0.12	2.8	2.8	2.8
	10	2.6	2.6	2.7
	20	2.4	2.3	2.6
	40	2.2	2.1	2.4
	70	1.9	1.9	2.2
A_{ef} (mT)		7.39	6.47	8.49

cay is considered to represent the decay process of ^{3,1}(K · · R) through reactions (5.28) and (5.30). The decay rates K_f in micellar SDS solutions of BP, BP-d_{10} and BP-^{13}C are given in Table 5.4.

From this table we see that there is a correlation between the absorption decay rates and the effective hyperfine coupling constants (last column) of the component radicals for the reactions of BP, BP-d_{10} and BP-^{13}C. This observation supports the idea that hf coupling is responsible for the variations of K_f in different isotope compositions of BP. The field dependence of the decay rate represents another manifestation of MIE. On the basis of these results, it was assumed [87] that the triplet-singlet conversion (reaction (5.28)) is the rate-determining step, so that the observed K_f was regarded as the rate constant of reaction (5.28). Qualitatively, this interpretation sounds correct. But quantitatively, there are some problems with it. The effective hyperfine coupling constants listed in the last column of Table 5.4 provide S-T mixing with a frequency about 10^7 s^{-1}, while the observed value of K_f is only about 10^6 s^{-1}, i.e., an order of magnitude less. Several explanations can be offered. One possibility is that not reaction (5.28), but rather reaction (5.30) is the rate-determining step. Another possibility is that fast recombination of RPs (reaction (5.30)) may reduce the rate of the singlet-triplet transitions due to a broadening of the RP singlet state energy level (see discussion of this problem in Chap. 3, Figs. 3.23 and 3.24). Consequently, K_f will be smaller than expected for the hyperfine interaction induced singlet-triplet transitions (reaction (5.28)) without taking into account this level broadening effect.

5.6 Isotope enrichment by resonant microwave pumping

The microwave field effect on radical reactions is a well established phenomenon (see, e.g., [47, 48, 63, 88–90]). Microwave pumping modifies the product yield of chemical reactions by inducing EPR transitions of the intermediate radical pairs. The microwave induced change of the product yield can be used as a method for isotope separation, since radicals with different isotope composition have different EPR spectra. For instance, the microwave field induced isotope enrichment in the photoreduction of menadione in micellar solutions was successfully observed by Okazaki et al. [91]. This reaction proceeds according to the following scheme [91]:

$$MD \longrightarrow {}^1MD^* \longrightarrow {}^3MD^* , \qquad (5.31)$$

$$^3MD^* + SDS \longrightarrow {}^3(MDH \cdot \cdot SDS) , \qquad (5.32)$$

$$^3(MDH \cdot \cdot SDS) \rightleftarrows {}^1(MDH \cdot \cdot SDS) , \qquad (5.33)$$

$$^3(MDH \cdot \cdot SDS) \longrightarrow MDH \cdot + \cdot SDS , \qquad (5.34)$$

$$^1(MDH \cdot \cdot SDS) \longrightarrow (\text{singlet product}) , \qquad (5.35)$$

$$\cdot SDS + TNO \longrightarrow T(SDS)NO \cdot . \qquad (5.36)$$

Menadione (MD) is promoted to its excited singlet state by UV light and converts to the lowest triplet state via intersystem crossing (reaction step (5.31)). Reaction step (5.32) describes the abstraction of hydrogen from one of the surrounding sodium dodecyl sulphate (SDS) molecules. Singlet-triplet transitions (reaction step (5.33)) enable the production of singlet molecules (reaction step (5.35)). Any radicals $\cdot$ SDS, escaping a cage reaction within micelles, are trapped outside the micelle (reaction step (5.36)). Perdeuterodimethylnitrosobenzene-sulphonate (TNO) serves as a trap molecule. The stable spin adduct T(SDS)NO $\cdot$ was detected by EPR technique. Photolysis was studied in normal SDS(H) micellar solution, in perdeuterated SDS(D) micellar solution and in micellar solution with a 1 : 1 mixture of SDS(H) and SDS(D). The microwave power was 5 mW, the frequency of the MW field was $5.939 \cdot 10^{10}$ rad/s. In this case, MW irradiation caused a decrease of the spin adduct yield by 15% for SDS(D) radical and by about 5% for the SDS(H) radical. These results are well understood in the framework of the MIE phenomenon for radical reactions. Microwave pumping selectively influences the singlet-triplet dynamics in RPs with different isotope composition (reaction step (5.33)).

6 Some perspectives

The magnetic isotope effect offers interesting aspects from many points of view.

Its basic property and foremost phenomenon is the dependence of chemical reactions on nuclear magnetic moments. Thus, MIE should be a subject of great interest to theoretical chemistry. MIE should also be taken into account in geochemistry [92] and cosmochemistry [93], since many oxidation-reduction processes in the Earth's core, photochemical processes in the atmosphere, hydrosphere and biosphere proceed via formation of radicals as intermediates.

MIE provides a novel mechanistic probe for geminate recombination of radical pairs. Several examples of the successful application of MIE for elucidating the routes of chemical transformation have been presented in Chap. 5. As a matter of fact, some interpretations of isotope effects in radical reactions, based on only the mass isotope effect might be reconsidered in view of the possibility of a magnetic isotope effect. It may well be that existence of a magnetic isotope effect will require changes of some accepted mechanisms of reactions.

Some initial results of magnetic isotope selection for heavy elements were presented in Chap. 5. An extremely interesting area of research in this field is the separation of nuclear isomers [94]. Nuclear isomers may differ by their spins and hyperfine coupling constants. For instance, the two isomers of tin, ^{119}Sn and ^{119m}Sn, have the nuclear spins 1/2 and 11/2, respectively. Therefore, these nuclear isomers may be separated in the course of radical reactions of elementoorganic compounds. The first attempt to separate the nuclear isomers of tin using MIE [94] was not successful. It could be attributed either to the large contribution of the spin-orbit coupling to S-T mixing in RPs or to the fast chemical exchange of Sn between reaction product molecules, but further research may open new gateways and provide new answers.

References

1. Melander L (1960) Isotope effects on reaction rates. Ronald Press, New York
2. Duncan JF, Cook GB (1968) Isotopes in chemistry. Clarendon Press, Oxford
3. Melander L, Saunders WH Jr (1980) Reaction rates of isotopic molecules. Wiley, New York
4. Lepley AR, Closs GL (eds) (1973) Chemically induced magnetic polarization. Wiley, New York
5. Salikhov KM, Molin YuN, Sagdeev RZ, Buchachenko AL (1984) Spin polarization and magnetic effects in radical reactions. Elsevier, Amsterdam; Akademiai Kiado, Budapest
6. Turro NJ, Kraeutler B (1984) Magnetic isotope effects. In: Buncel E, Lee CC (eds) Isotopes in organic chemistry, vol 6. Elsevier, Amsterdam, pp 107–161
7. Onsager L (1938) Initial recombination of ions. Phys Rev 54: 554–557
8. Rockwood S (1976) Uranium isotope separation and its demand on laser development. In: Mooradian A, Jaeger T, Stokseth P (eds) Tunable lasers and applications. Springer, Berlin, pp 140–149 (Springer series in optical sciences, vol 3)
9. Brocklehurst B (1969) Formation of excited states by recombining organic ions. Nature 221: 921–923
10. Lawler RG, Evans GT (1971) Chemical consequences of magnetic interactions in radical pairs. Ind Chim Belge 36: 1087–1089
11. Sagdeev RZ, Salikhov KM, Leshina TV, Kamkha MA, Shein SM, Molin YuN (1972) Influence of magnetic field on radical reactions. Pis'ma ZhETF 16: 599–602
12. Brocklehurst B, Dixon RS, Gardy EM, Lopata VJ, Quinn MJ, Singh A, Sargent FP (1974) The effect of a magnetic field on the singlet/triplet ratio in geminate ion recombination. Chem Phys Lett 28: 361–363
13. Schulten K, Staerk H, Weller A, Werner H-J, Nickel B (1976) Magnetic field dependence of the geminate recombination of radical ion pairs in polar solvents. Z Phys Chem N F 101: 371–390
14. Michel-Beyerle ME, Haberkorn R, Bube W, Steffens E, Schroeder H, Neusser HJ, Schlag EW, Seidlitz H (1976) Magnetic field modulation of geminate recombination of radical ions in a polar solvent. Chem Phys 17: 139–145
15. Buchachenko AL, Galimov EM, Ershov VV, Nikiforov GA, Pershin AD (1976) Enrichment of isotopes induced by magnetic interactions in chemical reactions. Dokl AN SSSR 228: 379–381
16. Molin YuN, Sagdeev RZ (1976) Spin effects in radical reactions. Lecture on all-Union conference on chemical kinetics dedicated to the 80th anniversary of academician N.N.Semenov. Moscow
17. Turro NJ, Kraeutler B (1978) Magnetic isotope and magnetic field effects on chemical reactions. Sunlight and soap for the efficient separation of ^{13}C and ^{12}C isotopes. J Amer Chem Soc 100: 7432–7434
18. Buchachenko AL, Khudyakov IV (1991) Magnetic and spin effects in photoreduction of uranyl salts. Acc Chem Res 24: 177–183
19. Nagakura S, Hayashi H (eds) (1987) Magnetic field effects upon dynamic behaviour and chemical reactions of excited molecules. Institute for Molecular Science, Japan
20. Kraeutler B, Turro NJ (1980) Probes for the micellar cage effect. The magnetic ^{13}C-isotope effect and a new cage product in the photolysis of dibenzyl ketone. Chem Phys Lett 70: 270–275

21. Sagdeev RZ, Leshina TV, Kamkha MA, Belchenko OI, Molin YuN, Rezvukhin AI (1977) A magnetic isotope effect in the triplet sensitized photolysis of dibenzoyl peroxide. Chem Phys Lett 48: 89–90

22. Tarasov VF (1980) Dynamics of the concentration of magnetic isotopes in chemical reactions. Zhur Fis Khim 54: 2438–2445

23. Rabinowitch E, Wood WC (1936) Collision mechanism and the primary photochemical process in solutions. Trans Far Soc 32: 1381–1387

24. Noyes RM (1955) Kinetics of competitive processes when reactive fragments are produced in pairs. J Amer Chem Soc 77: 2042–2045

25. McLauchlan KA, Steiner UE (1991) The spin-correlated radical pair as a reaction intermediate. Mol Phys 73: 241–263

26. Turro NJ, Weed GC (1983) Micellar systems as "supercages" for reactions of geminate radical pairs. Magnetic effects. J Amer Chem Soc 105: 1861–1868

27. Carrington A, McLachlan AD (1967) Introduction to magnetic resonance. Harper and Row, New York

28. Atherton NM (1973) Electron spin resonance. Wiley, New York

29. Abragam A (1961) The principles of nuclear magnetism. Clarendon Press, Oxford

30. Hellwege K-H, Hellwege AM (eds), Fischer H (1965) Magnetic properties of free radicals. Springer, Berlin [Mandelung O et al (eds) Landolt-Börnstein: numerical data and functional relationships in science and technology, NS, group 2, atomic and molecular physics, vol 1]

31. Harris RK, Mann BE (1978) NMR and the periodic table. Academic Press, New York

32. Landau LD, Lifshitz EM (1977) Quantum mechanics. Pergamon Press, Oxford

33. Schulten K, Wolynes PG (1978) Semiclassical description of electron spin motion in radicals including the effect of electron hopping. J Chem Phys 68: 3292–3297

34. Lueders K, Salikhov KM (1987) Theoretical treatment of the recombination probability of radical pairs with consideration of singlet-triplet transitions induced by paramagnetic relaxation. Chem Phys 117: 113–131

35. Salikhov KM (1983) On the largest possible contribution from hyperfine interactions to the recombination probability of radical pairs. Chem Phys 82: 163–169

36. Tomkiewicz M, Groen A, Cocivera M (1972) Nuclear spin polarization during the photolysis of di-t-butyl ketone. J Chem Phys 56: 5850–5857

37. Lueders K, Salikhov KM (1988) Theoretical treatment of CIDNP caused by anisotropic magnetic interactions. Chem Phys 128: 395–411

38. Salikhov KM, Mikhailov SA (1983) Calculations of the effect of magnetic nuclei on the recombination of radicals in the Earth's magnetic field. Teoret Exp Khim 19: 550–555

39. Salikhov KM (1983) Theory of magnetic effects in radical reactions at zero field. Chem Phys 82: 145–162

40. Deutch JM (1972) Theory of chemically induced dynamic polarization in thin films. J Chem Phys 56: 6076–6081

41. Fischer H (1983) The effect of a magnetic field on the product yield of a geminate radical pair reaction in homogeneous solution. Chem Phys Lett 100: 255–258

42. Tarasov VF, Pershin AD, Buchachenko AL (1980) ^{13}C enrichment in the photolysis of dibenzyl ketone in viscous solutions. Izv AN SSSR, Ser Khim 8: 1927–1928

43. Hayashi H, Nagakura S (1984) Theoretical study of relaxation mechanism in magnetic field effects on chemical reactions. Bull Chem Soc Jpn 57: 322–328

44. Shkrob LA, Tarasov VF (1990) Magnetic field and magnetic isotope effects in micelles. The kinetics of photochemical reactions. Chem Phys 147: 369–375

45. Steiner UE, Wu JQ (1992) Electron spin relaxation of photochemically generated radical pairs diffusing in micellar supercages. Chem Phys 162: 53–67

46. Nagakura S, Hayashi H (1983) External magnetic field effects upon photochemical reactions in solutions. Radiat Phys Chem 21: 91–94

47. Frankevich EL, Pristupa AI (1976) Magnetic resonance of excited charge-transfer complexes registered by fluorescence at room temperature. Pis'ma ZhETF 24: 397–400

48. Molin YuN, Anisimov OA, Grigoryants VM, Molchanov VK, Salikhov KM (1980) Optical detection of ESR spectra of short-lived ion-radical pairs produced in solution by ionizing radiation. J Phys Chem 84: 1853–1856

49. Kubarev SI, Pschenichnov EA (1974) The effect of high frequency magnetic fields on the recombination of radicals. Chem Phys Lett 28: 66–67
50. Salikhov KM, Molin YuN (1993) Some peculiarities of spin dynamics of geminate radical pairs under microwave pumping. J Phys Chem 97: 13259–13266
51. Sterna L, Ronis D, Wolfe S, Pines A (1980) Viscosity and temperature dependence of the magnetic isotope effect. J Chem Phys 73: 5493–5499
52. Turro NJ, Chung C-J, Jones G, Becker WG (1982) Photochemistry in ultrahigh laboratory magnetic fields. Photolysis of micellar solutions of dibenzyl ketones and phenylbenzyl ketones at 145000 G. Observation of $\Delta g\bar{H}$ effect on the cage reaction. J Phys Chem 86: 3677–3679
53. Turro NJ, Chow M-F, Kraeutler B (1980) The dynamics of the photodecarbonylation of dibenzyl ketone in a micellar detergent solution: effect of temperature on the absolute quantum yields and on ^{13}C enrichment. Chem Phys Lett 73: 545–549
54. Turro NJ, Kraeutler B, Anderson DR (1979) Magnetic and micellar effects on photoreactions. Micellar cage and magnetic isotope effects on quantum yields. Correlation of ^{13}C enrichment parameters with quantum yield measurements. J Amer Chem Soc 101: 7435–7437
55. Turro NJ, Chow M-F, Chung C-J, Weed GC, Kraeutler B (1980) Magnetic field and magnetic isotope effects on cage reactions in micellar solutions. J Amer Chem Soc 102: 4843–4845
56. Turro NJ, Chow M-F, Chung C-J, Smith WJ (1982) Magnetic and micellar effects in photoreactions. ^{13}C NMR determination of selective ^{13}C enrichment in a dibenzyl ketone photoproduct. Tetrahedron Lett 23: 3223–3226
57. Turro NJ, Chow M-F (1980) Magnetic isotope effect on the thermolysis of 9,10-diphenylanthracene endoperoxide as a means of separation of ^{17}O from ^{16}O and ^{18}O. J Amer Chem Soc 102: 1190–1192
58. Turro NJ, Chow M-F, Chung C-J, Tung C-H (1983) Magnetic field and magnetic isotope effects on photoinduced emulsion polymerization. J Amer Chem Soc 105: 1572–1577
59. Brocklehurst B (1974) Yields of excited states from geminate recombination of hydrocarbon radical ions. Chem Phys Lett 28: 357–360
60. Brocklehurst B (1976) Magnetic field effect on the pulse shape of scintillations due to geminate recombination of ion pairs. Chem Phys Lett 44: 245–248
61. Brocklehurst B (1983) Spin correlation effects in radiolysis. Radiat Phys Chem 21: 57–66
62. Dixon RS, Sargent FP, Lopata VJ, Gardy EM (1977) Fluorescence from γ-irradiated solutions of naphthalene and naphthalene-d_8. Effect of an applied magnetic field. Chem Phys Lett 47: 108–112
63. Anisimov OA, Grigoryants VM, Molchanov VM, Molin YuN (1979) Optical detection of ESR absorption of short-lived ion-radical pairs produced in solution by ionizing radiation. Chem Phys Lett 66: 265–268
64. Khudyakov IV, Serebrennikov YA, Turro NJ (1993) Spin-orbit coupling in free radical reactions. On the way to heavy elements. Chem Rev 93: 537–570
65. Groff RP, Suna A, Avakian P, Merrifield RE (1974) Magnetic hyperfine modulation of dye-sensitized delayed fluorescence in organic crystals. Phys Rev B9: 2655–2660
66. Bube W, Michel-Beyerle ME, Haberkorn R, Steffens E (1977) Sensitized charge carrier injection into organic crystals studied by isotope effects in weak magnetic fields. Chem Phys Lett 50: 389–393
67. Buchachenko AL, Fedorov AV, Yashina LL, Galimov EM (1984) The magnetic isotope effect and oxygen isotope selection in chain processes of polymer oxidation. Chem Phys Lett 103: 405–407
68. Yashina LL, Buchachenko AL (1990) Magnetic isotope effect and oxygen isotope selection in oxidation chain reactions. Chem Phys 146: 225–229
69. Turro NJ, Chow M-F, Rigaudy J (1981) Mechanism of thermolysis of endoperoxides of aromatic compounds. Activation parameters, magnetic field, and magnetic isotope effects. J Amer Chem Soc 103: 7218–7224
70. Step EN, Tarasov VF, Buchachenko AL (1988) ^{29}Si magnetic isotope effects in the photolysis of silyl ketones. Chem Phys Lett 144: 523–526

71. Wakasa M, Sakaguchi Y, Hayashi H (1992) The first direct observation of magnetic field effects on the dynamic behaviour of radical pairs involving group 14 silicon and germanium centered radicals. J Amer Chem Soc 114: 8171–8176

72. Hayashi H, Sakaguchi Y, Tsunnooka M, Yanagi H, Tanaka M (1987) Laser flash photolysis study of magnetic field effects on the photodecomposition of phenacyl phenyl sulfone in micellar solution. Chem Phys Lett 136: 436–440

73. Step EN, Buchachenko AL, Turro NJ (1992) Magnetic effects in the photolysis of micellar solutions of phenacylphenyl-sulfone. Chem Phys 162: 189–204

74. Taraban MB, Leshina TV, Salikhov KM, Sagdeev RZ, Molin YuN, Margorskaya OI, Vyasankin NS (1983) The influence of the magnetic field as a tool for investigation of the mechanism of reactions of triethylgermyl radicals. J Organomet Chem 256: 31–36

75. Wakasa M, Hayashi H, Kobayashi T, Takada T (1993) Enrichment of germanium-73 with the magnetic isotope effect. J Phys Chem 97: 13444–13446

76. Hayashi H, Wakasa M, Sakaguchi Y (1995) Magnetic field and magnetic isotope effects upon reactions of heavy atom-centered radicals. J Chinese Chem Soc 42: 343–352

77. Nikitenko SI, Gai AP, Glazunov MP, Krot NN (1990) Anomalous kinetic isotope effect in reaction of photochemical reduction of uranyl in oxalate solutions. Dokl AN SSSR 312: 402–406

78. Buchachenko AL, Khudyakov IV, Klimchuk ES, Golubkova NA (1990) Magnetic isotope effect in the photochemical reduction of uranyl nitrate. Izv AN SSSR, Ser Khim 8: 1902–1903

79. Closs G, Doubleday C Jr (1973) Determination of the average singlet-triplet splitting in biradicals by measurement of the magnetic field dependence of CIDNP. J Amer Chem Soc 95: 2735–2736

80. Closs GL, Redwine OD (1985) Cyclization and disproportionation kinetics of triplet generated, medium chain length, localized biradicals measured by time-resolved CIDNP. J Amer Chem Soc 107: 4543–4544

81. De Kanter F, Kaptein R (1982) CIDNP and triplet-state reactivity of biradicals. J Amer Chem Soc 104: 4759–4766

82. Zimmt MB, Doubleday C Jr, Turro NJ (1986) On the rate determining step for decay of triplet biradicals: intersystem crossing vs chain dynamics. J Amer Chem Soc 108: 3618–3620

83. Zimmt MB, Doubleday C Jr, Turro NJ (1987) Substituent and solvent effects on the lifetimes of hydrocarbon-based biradicals. Chem Phys Lett 134: 549–552

84. Turro NJ, Doubleday C Jr, Hwang K-C, Cheng C-C, Fehlner JR (1987) Magnetic isotope effects in biradicals: substantial ^{13}C enrichment and a novel mechanistic probe. Tetrahedron Lett 28: 2929–2932

85. Klein J, Voltz R (1976) Time-resolved optical detection of coherent spin motion for organic-radical-ion pairs in solution. Phys Rev Lett 36: 1214–1217

86. Veselov AV, Melekhov VI, Anisimov OA, Molin YuN (1987) The induction of quantum beats by the Δg mechanism in radical ion pair recombination. Chem Phys Lett 136: 263–266

87. Sakaguchi Y, Hayashi H, Nagakura S (1982) Laser-photolysis study of the external magnetic field effect upon the photochemical processes of carbonyl compounds in micelles. J Phys Chem 86: 3177–3184

88. Trifunac AD, Smith JP (1980) Optically detected time resolved EPR of radical ion pairs in pulse radiolysis of liquids. Chem Phys Lett 73: 94–97

89. Batchelor SN, McLauchlan KA, Shkrob IA (1992) Reaction yield detected magnetic resonance in exciplex systems. II. Time resolved and pulse studies. Mol Phys 75: 531–561

90. Okazaki M, Sakata S, Konaka R, Shiga T (1987) Product yield-detected ESR on magnetic field-dependent photoreduction of quinones in SDS micellar solution. J Chem Phys 86: 6792–6800

91. Okazaki M, Shiga T, Sakata S, Konaka R, Toriyama K (1988) Isotope enrichment by electron spin resonance transitions of the intermediate. J Phys Chem 92: 1402–1404

92. Galimov EM (1979) Nuclear-spin isotope effect – new type of isotope effect. Geo-khimia 2: 274–284
93. Haberkorn R, Michel-Beyerle M, Michel KW (1977) Isotope effects in interstellar molecules by chemical hyperfine interaction. Astron Astrophys 55: 315–318
94. Podoplelov AV, Leshina TV, Sagdeev RZ, Molin YuN, Gol'danskii VI (1979) Use of magnetic isotope effect to separate heavy isotopes, e.g., stannum. Pis'ma ZhETF 29: 419–421

Subject index

*Springer-Verlag
and the Environment*

WE AT SPRINGER-VERLAG FIRMLY BELIEVE THAT AN
international science publisher has a special obliga-
tion to the environment, and our corporate policies
consistently reflect this conviction.

WE ALSO EXPECT OUR BUSINESS PARTNERS – PRINTERS,
paper mills, packaging manufacturers, etc. – to commit
themselves to using environmentally friendly mate-
rials and production processes.

THE PAPER IN THIS BOOK IS MADE FROM NO-CHLORINE
pulp and is acid free, in conformance with international standards for paper permanency.